CONTENTS

HUMAN ORIGINS
THE SEARCH FOR OUR BEGINNINGS

Herbert Thomas

DISCOVERIES

HARRY N. ABRAMS, INC., PUBLISHERS

Six billion years from now, astrophysicists predict, the sun will have burned all its hydrogen. It will then expand and turn into a giant red star that will swallow up the earth. If human beings have not colonized another planet in our galaxy before then, they will clearly be doomed. When seen in the light of these billions of years, humankind's history has only just begun.

CHAPTER I
ANTEDILUVIAN PEOPLE

The Bible says that God made the world from the primitive chaos and in six days created the heavens, the earth and the waters, the fish, the birds, all the animals, and finally the first humans, Adam (right) and Eve. Not until the 20th century were humans able to contemplate the earth from space (opposite).

On the Sixth Day, "God Created Man in His Own Image"

With the emergence of monotheistic thought in the Judeo-Christian tradition, the biblical account of Genesis was for many years the sole explanation as to how the entire universe had been created. The human species, like all other animals, was formed as part of a grand design, conceived by the creator. It was not until the 19th century that a different interpretation was outlined, drawn from the observation of nature. Following several revolutions in the life and earth sciences, human beings were no longer considered to be the outcome of a unique and uniform creation in which species were fixed, but rather part of a long history in which all species evolve through natural selection.

Humankind and the present-day great apes have a common ancestor. This was the

During the 16th and 17th centuries in Europe prehistoric archaeology was still in limbo, despite the fascination that such megalithic monuments as Stonehenge (above) held for scholars and antiquarians.

In the Middle Ages the Bible provided the only explanation for the creation of the universe. This illumination shows the alpha and the omega, the first and the last.

19th century's shocking idea—a decisive concept if ever there was one, carrying with it the seed of the notion that modern biological science would from then on strive to demonstrate: that humans and our cousins the apes share similar features, just as they shared the same ancestor.

Today we know that almost ninety-nine percent of human and chimpanzee genetic material is the same. From the remaining one percent, we can reconstruct the gigantic puzzle of the history of our origins.

Humans Meet Apes

Around 460 BC the Carthaginian navigator Hanno sailed beyond the Pillars of Hercules (today's Strait of Gibraltar) at the head of sixty galleys with fifty rowers, to explore the west coast of Africa. Penetrating the interior of these lands, the soldiers saw strange hairy beings with human heads. Three of these creatures, which they called gorillas—they were probably chimpanzees— were captured and their skins were brought back to Carthage; at the time of the Roman conquest, in 146 BC, they could still be admired there, as Pliny tells us, in the temple of the goddess Astarte.

Regardless of how much truth there may be in this tale, three centuries later Galen, a Greek physician and anatomist (AD 129–c. 199), carried out the first autopsy of monkeys—macaques and baboons—which he considered to be "comical copies" of people. Thus, although these

This engraving, taken from the 1732 Bible of Canon Johann Jakob Scheuchzer, evokes in allegorical fashion the traditional vision of the earthly paradise and the creation, a literal expression of the holy text of Genesis.

hairy, tailed creatures were barely known—except, of course, Hanno's "gorillas"—they were already a source of fascination and a reason for concern.

Faced with this "family likeness," people were not interested in whether or not humans were descended from apes; they wanted to know whether apes were a type of human. This idea persisted until the second half of the 18th century, when the French philosopher Jean-Jacques Rousseau (1712–78) devoted long commentaries to it. Although denying a close relationship with chimpanzees—which are our closest relatives—he was uncertain about the position of the orangutan, discovered less than a century earlier. He wrote, "Perhaps after more precise research they will be found to be men." In the same century Georges Buffon, the famous French naturalist (1707–88), finally put an end to these "philosophers' daydreams" by contrasting the anatomy of the most human of the apes—the orangutan—with that of the most apelike humans, represented, according to him, by the Hottentots of southern Africa.

The Path Leading to the Theory of Evolution

Before the discovery of the first human fossils in the early 19th century, theories about human origins developed in a chaotic way with no logical thread linking them. Freethinkers, philosophers, naturalists, scholars, and antiquarians introduced various innovative ideas, which, of necessity, always

In 1658 Jacob de Bondt, a Dutch doctor who had lived in Java, was the first to describe the orangutan (left): "The Javanese give them the name of Orang Outang, which means people of the woods, and claim that they are born of the lewdness of Indian women who copulate with monkeys without tails."

St. Jerome points to a human skull, perhaps meditating on human nature or its destiny. This late-15th-century painting no doubt expresses the concerns of Renaissance thinkers rather than those of the 4th-century saint, who is known especially for his translations of the Bible and his exegetical treatises.

Before Buffon, the Hottentots—and the apes—were considered by some to be degenerate children of Adam. In 1829 a Hottentot woman, a Venus with very marked steatopygia (illustration below), was brought to Paris and put on exhibit.

went against biblical tradition; they were very speculative in nature.

While, on the one hand, Irish Archbishop James Ussher (1581–1656)—basing his figures on biblical chronology—had calculated that the world was created in 4004 BC, there is evidence, on the other hand, that a variety of thinkers from the 15th to the 18th century came to other conclusions. Such figures as Leonardo da Vinci, French poet Cyrano de Bergerac, German philosopher Gottfried Wilhelm Leibniz, and Georges Buffon were thinking in terms of much longer timespans and considering the possibility that the universe had been in existence for hundreds of thousands—even millions—of years, though they were not able to prove the fact.

In 1616 Lucilio Vanini (1584–1619), an Italian philosopher, suggested that humans originated from the ape, an idea for which he was burned alive three years later in Toulouse. In the following century the French physician and philosopher Julien Offroy de La Mettrie (1709–51) did his utmost to show that the differences between humans and the "lower echelons"—including plants—are only a question of degree.

In his 1731 work *Physica Sacra* Scheuchzer depicted the "witness of the Flood" (opposite) and other human remains, like these two vertebrae (left) collected near Altdorf, Germany.

In fact, the fundamental idea that humans might be descended from apes did not really take shape until Charles Darwin (1809–82) published his theory of evolution in *The Origin of Species* (1859). Species evolved from other species, he said, through natural selection.

Homo Diluvii Testis

In the period when these theories were being developed, numerous fossil bones—which were then called "petrifications"—were frequently unearthed in different parts of Europe. At the start of the Age of Enlightenment, Canon Johann Jakob Scheuchzer (1672–1733), a Swiss doctor and, moreover, a zealous propagandist for the popular biblical Flood theory, made a remarkable discovery. He found the imprint of a "human skeleton" on a slab of schist extracted from a quarry at Oeningen, near Lake Constance. Described as one of the "rarest relics of the accursed race that must have been swallowed up by the waters," the find appears under the name of *Homo diluvii testis* ("witness of the Flood") in a very curious work by Scheuchzer on fossil fish, called *Piscium Querelae et Vindiciae*, published in Zurich in 1708.

The canon stuck to his ideas about the Flood, and a few years later he made a second discovery—this time of some black, shiny fossil vertebrae from the foot of a gallows near Altdorf, Germany. Although some of the more astute naturalists of the day expressed doubts about Scheuchzer's findings, almost a century was to

pass before the true nature of these fossils was at last recognized.

In 1811 Georges Cuvier (1769–1832), the great French naturalist and father of vertebrate paleontology, was passing through Haarlem, in Holland, where the famous Oeningen skeleton was kept. In front of witnesses Cuvier demonstrated conclusively that the bones from the supposed "witness of the Flood" in fact came from a giant salamander. It is hard to believe that Scheuchzer, being a doctor, could have confused a giant salamander skeleton with one from a human!

As for the vertebrae, Cuvier, who had seen similar specimens in the Grand Duke of Tuscany's natural history collection, attributed them quite correctly to an extinct marine reptile, the ichthyosaur, the existence of which had just been discovered by English scholars.

In Cuvier's day the hill of Montmartre in Paris was a major center for the mining of gypsum, a rock used in making plaster. The gypsum, deposited there in the late Eocene period, more than forty million years ago, contains numerous fossil bones that workers collected and Cuvier reconstructed.

Starting in 1796 Cuvier (left) studied the numerous remains of fossil vertebrates that were known at the time. Basing his theory on one of the fundamental laws of comparative anatomy, which he had discovered, Cuvier set about reconstructing the animal's entire skeleton on the basis of sometimes fragmentary evidence in his famous work *Recherches sur les Ossements Fossiles de Quadrupèdes* (1812).

No fossil primates had been found by the beginning of the 19th century. It is to Cuvier that we owe the description in 1822 of the first one (below), based on a small jawbone discovered in the gypsum quarries of Montmartre. It belonged not to an ape but to an animal form close to present-day lemurs.

"There Are No Human Fossils" (Cuvier)

In the first half of the 19th century, Cuvier's prestige was considerable. His revelation that strange animals—which Cuvier had reconstructed from fossil bones found in the gypsum quarries of Montmartre—had once inhabited the site of Paris itself made him a kind of god in the eyes of the Parisian public and his contemporaries. In 1831, just before the illustrious scientist died, the French novelist Honoré de Balzac captured the mood of the time and hailed Cuvier as the greatest poet of his century in a lyrical

A

A. STEER ATTENTION to important matters w[...]
realistically dimensional, hand-painted poly resi[...]
magnets, one each: Block Island, RI; Yaquina[...]
and West Quoddy, ME. Widest, 3-1/4"; [...]
#186684 - set of three lighthouse mag[...]

B. YOU'RE SHIPSHAPE on the [...]
ahead with our afghan on deck [...]
woven in South Carolina usi[...]
washable cotton. 46" [...]
max. 16 letter[...]
**#31067[...]
**#85[...]

TSIDE?
can see
t with
ttom that
patina.

G

F

voodlily
fted
nnecticut-
is fused
then
lass

G

Gardeners
are a
Gift

passage. Of course, the century was only thirty years old at that point. ...

Despite his scientific clairvoyance, Cuvier rejected the theory of evolution, clinging to the biblical Flood theory, and he set about explaining the appearance of new species according to his belief in catastrophic destruction and the immutability of species. He brought all of his considerable authority to bear on the delicate question of early human remains, whose existence he denied—though he was careful to add that they did not occur "at least in our countries."

It is true that Cuvier relied on arguments that were believed to be a priori legitimate at this time. First, he based his theory on the fact that he did not find any human or ape bones among the skeletons of extinct animals recovered from the Montmartre gypsum quarries. (Naturally he did not know that the gypsum deposits imprisoning all these animals dated back more than forty million years—so these animals lived long before either humans or even apes had appeared.) Second, it seems that all of the many skeletons that had been presented to Cuvier in the belief that they were antediluvian people belonged, in fact, to elephants, tortoises, salamanders, cetaceans, or marine reptiles.

In 1735, in his *Systema Naturae*, Linnaeus (below, with a section of his seminal work) proposed a system of binary nomenclature for classifying living beings. He felt it accounted for the plan of the divine creation. In his *Fundamenta Fructificationis* of 1762, however, he admitted that there may be a common root for all the species in the same genus, if not in the same order. Thus, God's work would have stopped with the genera or orders, their further diversification being attributed to crossbreeding or hybridizations.

ANTHROPOMORPHA.

Dentes incisores IV. *supra & infra.* *Mammæ pectorales.*

1. HOMO Nosce te ipsum. *
  1. Homo, *cujus varietates* Europæus albus. - - - *l'Homme.*
                    Americanus *rufescens.*
                    Asiaticus *fuscus.*
                    Africanus *niger.*

2. SIMIA. *Facies* nuda. *Ungues* planiuscu
            tundati. *Cilia* utrinque.

  1. Simia ecaudata subtus glabra. *Tulp. obf. t.* 271. *Bont. ind.* 8...
  2. Simia ecaudata, unguibus indicis tubulatis. *Seb. thef.* 1. ...
  3. Simia ecaudata, clunibus tuberosis. *Alp. ægypt.* 241. *t.* 1...
  4. Simia ecaudata, rufo nigricans. *Alp. ægypt.* 242. t. 20. *f.*...
  5. Simia femicaudata, ore vibriffato, unguibus acutis. Papio. *R...*
  6. Simia caudata, ore ciliisque vibriffatis. *Pet. gaz. t.* 13...
  7. Simia caudata imberbis, auribus comofis. *Alp. ægypt.* 242...
  8. Simia caudata imberbis, unguibus pollicum fubrotundis...
  9. Simia caudata imberbis fufco-flava, pectore gulaque al...
                    cofa. *Cluf. exot. t.* 371.
  10. Simia caudata imberbis, naribus elatis bifidis. *Marc...*
  11. Simia caudata imberbis, collo pectoreque jubatis.
  12. Simia caudata, genubus auribusque barbatis. *Mar...*
  13. Simia caudata, cæfarie prolixa faciem cingente. ...
  14. Simia caudata barbata, cauda prehenfili. *Marcgr.*
  15. Simia caudata barbata, cauda floccofa. *Cluf. exot.*
  16. Simia caudata barbata, barba cana, cauda fimplic...

Linnaeus' Daring Idea: Classifying Humans Close to the Great Apes

At the same time research was continuing into the classification of animals. In 1735 Swedish botanist Carolus Linnaeus (1707–78) published his *Systema Naturae*. In this system, which is still in force today, every living form is designated by two Latin names—a genus name followed by a species name. For example, in *Homo sapiens*, which means wise human being, *Homo* is the genus and *sapiens* the species. Since all people can interbreed, by definition they belong to the same species. (Of course, this criterion for identifying a species cannot be applied to fossils; other standards have to be used to identify and classify the remains of human species that preceded *Homo sapiens*.) Finally, the whole group of related species that share one or several evolved characteristics will belong to the same genus. Thus the dog, wolf, fox, and jackal belong to one and the same genus: *Canis*.

One of Linnaeus' daring ideas was to classify humans close to a few great apes, including the gibbon. He was not trying to express a family connection between humans and gibbons but to account for the divine creator's plan. The Swedish scholar called this group the primates, because in his view they held the first rank in nature's hierarchy; the other mammals were accorded the second rank, and the reptiles were relegated to third position: hence the names primates, secundates, and tertiates. Only

In the tenth edition of his *Systema Naturae* (1758), Carolus Linnaeus created the order primates as well as the genus *Homo*, which, he said, comprised two species: "day man" and "night man." The latter includes "nocturnal man," the "man of the woods," and the orangutan. Such distinctions indicate the uncertainty about and imprecision of the boundaries between apes and humans. Linnaeus himself admitted that he "[had] not up to now extracted from the principles of [his] science any characteristic that might make it possible to distinguish humans from apes."

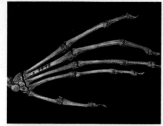

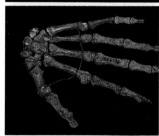

the first survives in modern taxonomy—the primates, the order that includes prosimians (the tarsiers, lorises, lemurs) and anthropoids (all New World and Old World monkeys and apes—as well as humans).

Classification: The History of Connections Between Living Beings

If we are to place humans within this order and study our origins, it is essential to understand a few principles of the modern language of classification.

Based on organic similarities, classification tries to account for the relationships between the numerous living forms. However, these similarities do not all have the same significance; indeed, some do not express any real relationship at all. For instance, the wings of a butterfly, a swallow, and a bat are organs that appeared independently in insects, birds, and mammals as an adaptation for flight. Other similarities are due to characteristics that have remained unchanged since their initial state. These features, known as primitive, are evidence of only a very distant connection. Our five fingers and toes, for example, relate back to the origin of the tetrapod—four-limbed—vertebrates.

Other characteristics—known as evolved or derived, as they are distant from the initial state—differentiate the primates, for example, from the other mammals: the growth of the brain, leading to the reduction of

Since Aristotle it has been observed that the feet and hands of nonhuman primates resemble each other, and the animals have been called quadrupeds or quadrumanes in turn. Whereas the hands of nonhuman primates have a double purpose— they are used for both locomotion and grasping —in humans only the latter function remains. What is particularly remarkable about the human thumb is that it can swivel fifty-four degrees and so can push against all the other fingers. In the indri, a large lemur from Madagascar, the thumb is separated from the other fingers (top); in the gorilla, which is more like us, the thumb can rotate (center). This movement attains its maximum range in humans (bottom).

the snout; the acquisition of three-dimensional vision and opposable thumbs; and the replacement of claws with flat nails.

As we move up the hierarchy of evolved characteristics in primates, we inevitably encounter characteristics that are exclusively human, such as our mode of locomotion—bipedal (on two legs)—or the growth in certain parts of the brain such as Broca's area (the motor speech center) that, together with the increase in volume of the pharynx and the lower position of the larynx, enabled us to acquire articulate speech and develop abstract thought.

It is clear that evolved characteristics play an important role, as they show family connections; primary characteristics, on the other hand, are merely an expression of at best a distant kinship and at worst a vague similarity.

The "Red Lady," the First Specimen of a Cro-Magnon

Until 1820 the rare discoveries of authentic human remains—such as those buried in a cave at Gailenreuth, southern Germany, or the worked flint axe found by the English antiquarian John Frere (1740–1807) in a Suffolk, England, quarry with the remains of extinct animals—did not lead anywhere, either because they went unnoticed or because they were not really understood.

However, after 1820 the number of excavations in caves started to increase rapidly. In 1822 the Reverend William

In depicting the fetus of a monkey and a human side by side (above) in 1812, one author wished to show the resemblance between them: "Except for the soul, it lacks nothing that we have."

William Buckland (left), professor of geology at Oxford, did not believe that humans and large extinct animals lived at the same time despite his own countless discoveries, such as the skeleton of the "Red Lady," which he thought had been buried during the Roman period.

Buckland (1784–1856) was digging in a cave called Paviland in Wales, and he unearthed a skeleton covered in ocher. It was not until many years later that the "Red Lady" was revealed to be the first specimen of a Cro-Magnon.

More discoveries soon followed. Excavations carried out by a pharmacist from Narbonne, France, called Paul Tournal (1805–72) revealed that human bones had been mixed with the remains of extinct animals in several caves in southern France. In 1830 the Belgian naturalist Philippe-Charles Schmerling (1791–1836) brought to light the first evidence of Neanderthals: the remains of a child's skull found near Liège, in eastern Belgium.

Despite the real antiquity of all these human remains, unfortunately none of them presented any particular anatomical traits that distinguished them immediately from modern humans. Consequently their immense significance was overlooked. Although the fields of archaeology and geology were increasingly overlapping, and it seemed obvious that fossil humans were contemporaneous with extinct animals, very few people reached the inevitable conclusion about the period in which humankind appeared.

It is true that Cuvier, who was still very influential, believed that humankind appeared at the same time as the ape. But no fossil apes had been found. Their discovery, which was to follow shortly, therefore indirectly contributed to the theory of human origins.

Pliopithecus Antiquus

One of the many people who attended public classes, especially those given by Cuvier at the Jardin des Plantes, in the Latin Quarter in Paris around 1820, was a young man by the name of Edouard

In 1833 Schmerling published an important work on fossil bones found in the Belgian province of Liège. It featured bones from extinct animals as well as humans, including this child's skull (above), which would not be recognized as Neanderthal and of great antiquity until 1936.

Left: An engraving from Buckland's *Reliquiae Diluvianae*, published in 1823, in which he described his excavations in England.

Lartet (1801–71). Although his father had sent him to Paris to study law in order to prepare him for a career, he was, in fact, to become one of the founders of human paleontology.

Once his studies were finished, Lartet returned to southwestern France, to his native Gascony, to settle down in the family home, the Château d'Ornezan. The windows of this old manor house look out at the hill of Sansan in the distance. By one of those quirks of fate that were so frequent in the 19th century, this hill would become one of the key sites in French paleontology.

In 1834 a shepherd brought Lartet a large tooth that he had found at the bottom of Sansan's slopes and told the young man that one could collect bones by the shovelful there. Lartet recognized that the tooth was from an extinct animal species—a mastodon—and, armed with spades and picks, he carried out major excavations and brought to light countless remains of fossil mammals. Among them, one lower jaw particularly attracted his attention—it came from the first fossil ape to be found. His discovery was announced in 1837.

Pliopithecus antiquus, as it would later be called, bore a strange resemblance to the modern gibbons of Southeast Asia. The eminent French naturalist Etienne Geoffroy St.-Hilaire (1772–1844), who had been alerted by Lartet, did

Ducrotay de Blainville, then professor of comparative anatomy at the Paris Muséum d'Histoire Naturelle, compared the first fossil ape (lower jaw, below) discovered by Lartet with the modern gibbon.

Edouard Lartet (left) was the first to propose a classification of Paleolithic times based on extinct mammals and the typological characteristics of the stone tools that accompanied them. Numerous sites discovered or reported by him have given their names to various cultures of the Middle and Upper Paleolithic.

not overlook the significance of this find, which, he claimed, was destined to "start a new era in humanitarian knowledge." Indeed, both Lartet and Geoffroy St.-Hilaire argued that fossil humans must exist because fossil apes did.

Some twenty years later, in 1856, Lartet had another stroke of good fortune. An amateur naturalist presented him with the lower jaw of another ape found in a tile works on the outskirts of St.-Gaudens, in southwestern France. The teeth of this new fossil ape, which Lartet called *Dryopithecus*, or ape of the oaks, were similar to those of modern great apes and even humans. The existence of such fossil apes turned the possibility of discovering fossil humans into a probability.

From Celtic Antiquities to Antediluvian People

Whereas Tournal, Schmerling, and a few others were certain that humans had existed in prehistoric times, official scientific thought had not made much progress. Somewhat unexpectedly, it was due to the stubbornness of an amateur, Jacques Boucher de Crèvecoeur de Perthes (1788–1868), that—

Ce fut en 1835 qu'en exam environnent Abbeville, pour y c Je pensai que s'il existait des ho for ma ces bancs, c'est là qu'on en trouver la trace...

after many trials and tribulations— the academic authorities of the period were finally convinced.

Boucher de Perthes, who came from a wealthy family, was the director of the customhouse at Abbeville, in northern France, and possessed literary

Several skeletons found on the island of Guadeloupe in 1803 (engraving, left), near the port of Le Moule, were first taken to be ancient human remains, but Cuvier said they were merely "corpses who perished in a few shipwrecks."

In a letter dated 1859 Boucher de Perthes (below) informed Isidore Geoffroy St.-Hilaire, the son of the French naturalist, of the circumstances that led him to discover "works by men in the Flood" in the period starting in 1835.

This stone tool, a biface, was found in the gravel quarries of St. Acheul (watercolors left), near Amiens, France.

talents that he used in various different genres. In addition, he was a great lover of antiquities—this was the period of Romanticism. He never tired of wandering the vast chalky plateau of Picardy, cut by the valley of the Somme, in search of worked stone tools. In the course of the years from 1835 to 1846, he collected more than a thousand worked flints from the ancient alluvium of the region's high terraces.

In 1847 he was finally in a position to begin to publish his efforts in a monumental work, *Antiquités Celtiques et Antédiluviennes*, in which he demonstrated in particular that humans had been the contemporaries

In 1863 Boucher de Perthes unearthed the jaw of an alleged antediluvian human fifteen feet below the ground in the Moulin-Quignon quarry near Abbeville (opposite below). (It had in fact been placed there fraudulently.) This discovery, which was the source of much debate among British and French scholars, would, paradoxically, bring him the fame to which he had aspired his whole life.

of the great antediluvian animals: the mammoth, the southern and straight-tusked elephants, the Etruscan ox, and the early bison and rhinoceros.

While retaining the traditional division of world history into two epochs separated by the universal catastrophe of the Flood, which he transposed into the domain of archaeology, Boucher de Perthes foreshadowed in the first volume of his great work the current subdivision of prehistoric times into Paleolithic (flaked stone culture) and Neolithic (polished stone culture), terms that would only be created some twenty years later by Sir John Lubbock (1834–1913).

Unfortunately, Boucher de Perthes' book also contained numerous flights of fancy of varying degrees of eccentricity, which were mingled with digressions on the origin of art. For instance, he claimed that some curiously shaped stones had been worked by people rather than nature, which was the actual case. The work was greeted with virtually unanimous skepticism, and added to the extreme reluctance of the French Académie des Sciences and the Académie des Inscriptions et Belles-Lettres to accept the new ideas.

A meeting of the Académie des Sciences is in session. In 1846 Boucher de Perthes addressed his unfinished report, *Antiquités Celtiques et Antédiluviennes* to this eminent body. Unfortunately, one of the members of the commission appointed to examine Boucher de Perthes' work completely rejected his convictions.

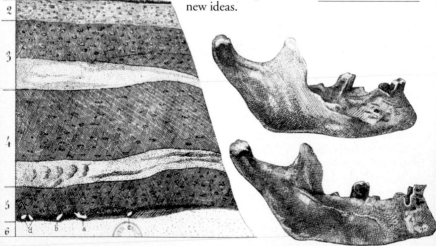

Far from being discouraged, Boucher de Perthes went right back to work and published the second volume of his *Antiquités Celtiques et Antédiluviennes* ten years later. In it he dealt with the objections to his first publication.

The Idea That Fossil Humans Were Contemporaries of Extinct Species Finally Found Acceptance

Although many scientists were still inclined to wait and see, the new thinking managed to win over some supporters, and prestigious ones at that, such as Edouard Lartet and Etienne Geoffroy St.-Hilaire.

From the mid-19th century on, events were to progress

very quickly. In 1859, the year in which it was founded, the Société d'Anthropologie in Paris devoted several of its sessions to the discoveries made by Boucher de Perthes.

In the same year British scientists, whose work played a large part in the establishment of the truth, gave their support to these new ideas. Neither the Scottish paleontologist Hugh Falconer (1808–65), the English archaeologist John Evans (1823–1908), nor the British geologists Joseph Prestwich (1812–96) or Charles Lyell (1797–1875) had any doubts that flint tools were made by humans who had lived at the same time as extinct animal species.

As for those with an unshakable belief in catastrophism —the last of those to share Cuvier's outmoded view— their days were numbered.

These engravings from the second half of the 19th century portray primitive humans grappling with ferocious beasts in a world filled with hostile forces. They were used to illustrate one of the popular works on science by the writer and doctor Louis Figuier. In his praiseworthy desire to instruct, Figuier denied the "distressing doctrine" extolled, he said, by the "materialist sect" that imagined a link between apes and humans. But Charles Lyell (below), the foremost British geologist, adopted Boucher de Perthes' ideas and would secure a place in history for the proof of human antiquity.

In 1911 amateur archaeologist Charles Dawson discovered a skull and a lower jaw of a fossil human in a gravel pit near Piltdown, in southern England; it became the subject of several hundred publications. Many scholars were convinced that the earliest-known human from western Europe had been found. In 1953 the truth was revealed: "Piltdown Man" was a fraud—the jaw of an ape had been combined with a human skull. If Charles Dawson was the instigator of this hoax, it was likely, given his lack of expertise, that he would have had an accomplice. Suspicion has fallen on different people. In *Piltdown: A Scientific Forgery* (1990), Frank Spencer names the famous English anatomist Sir Arthur Keith as the chief suspect. This 1915 painting of the people most involved in the "discovery" and inter-pretation of the "fossil" shows Keith, in a white jacket, examining the Piltdown skull. Standing behind, to his left, is the enigmatic Dawson.

1. Mittelländer (Kaukaſier). 2 Nubier. 3. Dravida. 4. Malaye. 5 Mongole. 6. Amerikaner.
7. Arktiker (Eskimo). 8 Auſtralier. 9. Neger. 10. Kaffer. 11. Hottentotte. 12. Papua.

The remains of prehistoric people and their tools began to be interpreted in the 19th century, as such new ideas as Darwin's theory of evolution emerged. It became apparent that humankind was closely related to the great apes.

CHAPTER II

HUMANS: NAKED APES OR LITTLE GODS?

The diversity of environments that humans have colonized during the last 100,000 years helps to explain their physical differences. Any division of humanity into races, as was popular in the 19th century (opposite), is, in reality, quite arbitrary. Humanity is the fruit of a single history, the dawn of which was shared with the great apes more than five million years ago.

Right: *The Travelled Monkey*, a 19th-century painting.

As the 19th century progressed, the belief in the biblical Flood theory gave way to a more scientific approach to our origins. Whatever the legitimacy of the metaphysical approach, it is now more than three centuries since people were burned for heresy, and the weakening of dogmas has contributed to a renewed fascination for the prehistory of humanity. As the stakes are high, one inevitable consequence is that old, theologically based ideas surface again periodically.

In actual fact, the close kinship of humans and great apes does not mean that humans are based on the same exact model as animals. The family relationship is certainly clear from the way both are classified according to common traits. However, this classification simply reflects our genealogy.

In the course of their evolution, humans became capable of communicating and developed a most important attribute: cultural transmission. In fact, in the words of Nobel Prize–winning cellular geneticist François Jacob, humans gradually became "programmed to learn," thus moving away from simple biological transformation, the program for which consisted almost entirely of copying innate behavior from one generation to another.

The different cultures —such as ancient Egyptian and medieval European (above)—that have succeeded each other have helped to shape our belief in the advancement and adaptability of the human race. Today the racist philosophies of the past have given way to the predominant view that our differences stem from cultural, not biological, variations. Thanks to human evolution and the passing of knowledge from generation to generation, we may soon—in the 21st century—be in a position to harness the energy of thermonuclear fission and colonize part of our solar system (opposite above).

Are Humans Part of the Animal Kingdom?

During the Second Empire (1852–70), the question of
whether humans belonged to the animal kingdom was
still regularly provoking agitation in the stormy meetings
of the French Société d'Anthropologie. Along with other
French naturalists, Jean-Louis Armand de Quatrefages
(1810–92), professor of anthropology at the Muséum
d'Histoire Naturelle and director of the anthropology
society, defended the existence of a human kingdom
distinct from the animal kingdom. He believed that his
view was justified by the fact that only humans possess
a notion of good and evil and a belief in an afterlife and
in superior beings.

Quatrefages was not an evolutionist, in the sense that
he did not believe that one species could descend from
another, and he could not imagine that the classification
of living beings might reflect their history. For him,
classification was merely a convenient contrivance
meant to smooth language and permit reasoning. He
felt that the creation of a category for humans alone,
separate from the animal kingdom, was inevitable.
Johann Friedrich Blumenbach, a German anthropologist
(1752–1840), shared the same anthropocentric view. In

In the course of history,
humans have gradually
been less influenced by
the natural world and the
changes in climate that
were responsible for their
evolution. Evolution's
changes, which become
apparent only with the
passing generations, take
place slowly because they
conform to the laws of
genetics. With the
appearance of articulate
language and then, much
later, of writing, people
have reached a new
evolutionary stage—a
cultural one—in which
every piece of acquired
knowledge is rapidly
transmitted in each
generation.

Première espèce humaine supérieure:
Méditerranéens (12), *quatre races*:
  12ᵃ *Sémites*, 12ᵇ *Basques*,
  12ᶜ *Caucasiens*, 12ᵈ *Indogermains*.

Cercle polaire
septentrional

Finnois

Scandinaves

Celtes

Basques

Berbères

Méditerranéens

Nègres de
Guinée

Hottentots

Madécasses

12	10	8	
Méditerranéens	Dravidiens	Arctiques	
11	9	7	
Nubiens	Américains	Mongols	Au

The eminent German
zoologist Ernst
Haeckel (1834–1919)
subdivided humanity
in his *Natürliche
Schöpfungsgeschichte*
(1868) into twelve species
on the basis of hair type,
skull shape, skin color,
and eye color, perhaps
using instruments like
this board of artificial
eyes (opposite) and this
palette of skin colors
(left). In Haeckel's
opinion, the twelve
human species had
arisen from one single
ancestral type, his *Homo*

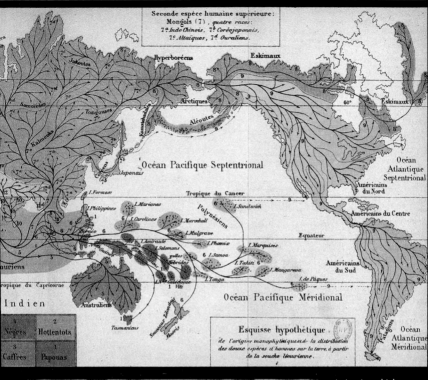

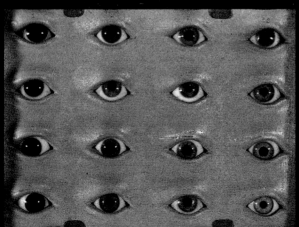

primigenius, which originated on a continent he imagined to be submerged beneath the present-day Indian Ocean, touching Madagascar and East Africa (in purple on the map).

1791 he had already contrasted bimanous beings (humans) with all the other animals, the quadrumanes. Others, such as German zoologist Johann Karl Illiger (1775–1813), had already mentioned the *Erecta* (named for their upright position), while English paleontologist Richard Owen (1804–92) was talking in terms of beings with superior brains. Today, however, nobody seriously doubts that humans form part of the animal kingdom.

An Unwanted Ancestor

In 1809 the French scientist Jean-Baptiste de Monet de Lamarck (1744–1829) published his *Philosophie Zoologique*, a prophetic book in which he expounded

Darwin avait raison

Roman ciné par MAURICE AUBYN FOX-FYLM

This typical early-19th-century anthropological instrument was used to measure human skulls, whose measurements, it was believed, revealed racial characteristics.

In the 1920s, Protestant traditionalists campaigned against the "antibiblical" evolutionary ideas then in vogue in American school textbooks. Not wishing his or anybody else's children to be forced to study such books, Tennessee legislator John Washington Butler managed to have a law passed in 1925 forbidding the teaching of the theory of evolution in public schools.

his theory of evolution and set out a system to explain the acquisition of new traits. However, even though he had toned down his evolutionary ideas, for fear of going too far, they did not find acceptance at the time. It should be remembered that, in this period, the emperor himself was using his despotic authority to uphold the highly popular creationist views that Cuvier championed.

Half a century later, in 1860, English biologist Thomas Huxley (1825–95), nicknamed "Darwin's bulldog," gave a brilliant lecture setting out his illustrious colleague's ideas and incurred the wrath of Bishop Wilberforce of Oxford, who was present. When the bishop asked him whether he claimed descent from an ape through his grandfather or his grandmother, Huxley retorted scathingly that he would rather have an honest ape for his grandfather than a man of restless and versatile intellect who used skills of oratory to obscure scientific questions of which he knew nothing.

Like the bishop of Oxford, most people still found the idea that they descended from apes profoundly humiliating.

At the famous Scopes ("Monkey") trial, in Tennessee in 1925, evolutionists clashed with the supporters of divine creation. Above: The headquarters of the Anti-Evolution League, which stood opposite the courthouse in which John Thomas Scopes, a science teacher, was sentenced to pay a $100 fine after having been found guilty of teaching that humans are descended from an inferior order of animals. The trial pitted defense attorney Clarence Darrow against prosecutor William Jennings Bryan.

A Treasure in a Cave

The year 1856 saw the discovery not only of the first fossil apes but also of the first recognized fossil humans. In August workers quarrying marble uncovered a little cave in a vertical cliff about sixty-five feet high, dominating the Neander Valley, near Düsseldorf. Beneath their feet the quarrymen were surprised to find a human skeleton in the thick layer of clay; it was stretched out, with its head turned towards the cave entrance.

Believing they had discovered the remains of a cave bear, which were often found in neighboring caves, the workers did not take any special care of the bones and in fact discarded them. Luckily, the quarry's director saw fit to notify one of his friends, a natural-history teacher named Fuhlrott, who taught nearby, in the village of Elberfeld. Fuhlrott immediately recognized that the remains came from a human being. He thus saved from complete destruction a skullcap, the femurs, humeri, ulnas, a clavicle, a shoulder blade, half a pelvis, and a few ribs.

Fuhlrott was struck by the skullcap's primitive appearance, with its very low vault, receding forehead, and, especially, enormous brow ridges. He was also intrigued by the unusual thickness of the walls of the bones, which led him to think that the individual to whom these remains belonged was extremely muscular.

The Neanderthal skullcap, above, which was discovered in 1856 provoked thirty years of memorable debates.

Fuhlrott remembered that the presence of enormous brow ridges was a feature that the English scientist Owen had pointed out in the gorilla, in 1848, on the basis of two heads sent to him by a Protestant missionary in Gabon. Fuhlrott was convinced that he had in his possession the remains of a remote ancestor, an intermediary between the great apes and humans.

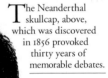

The German anatomist Schaaffhausen reconstructed the image of the Neanderthals in 1888 (above) from two skeletons discovered in a cave at Spy, in Belgium, two years earlier.

Neanderthal: A Controversy Lasting Several Decades

As the skull had been found by mere workmen, some people thought they could dispute its authenticity. The question of the primitive Neanderthals was fiercely debated at the Kassel congress of 1857, where the most improbable flights of fancy were constructed. According to one scholar, the skull probably belonged to an unintelligent hermit. For the French scientist Franz Pruner-Bey, however, it was without doubt a representative of the Celtic race. Others thought they were dealing with a Russian cossack who had died there in 1814; it was opportunely recalled that General Chernichev's cossack army had indeed camped nearby.

Only the anatomist Hermann Schaaffhausen (1816–93) from Bonn shared Fuhlrott's point of view. He felt that the human skull that he held in his hands was by far the most interesting he had seen in his entire life. For him, the Neanderthal remains undeniably displayed apelike features. In fact, Schaaffhausen, who had long supported evolution, was convinced that these remains provided proof of humankind's animal origins.

Everyone waited for an opinion from the famous pathologist Rudolf Virchow (1821–1902), who was an undisputed authority in his field in Germany at the time. Virchow's pronouncement was that it was not

Nowadays Neanderthals are considered to be very close to modern humans; indeed they are classified as *Homo sapiens neanderthalensis*. These stocky and very muscular people had large flattened heads with prominent noses and receding cheekbones. Below: A late-19th-century reconstruction of a Neanderthal man.

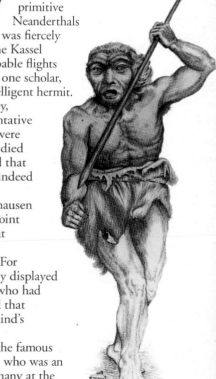

from a normal man but a malformed idiot who had been afflicted with rickets and arthritis.

Charles Lyell, having carefully examined the remains as well as the cave, felt it represented a new species of human distinct from *Homo sapiens*. His assistant, William King, even named it *Homo neanderthalensis* in 1864. In fact, the controversy raged on for almost three decades.

With the discovery of the Neanderthal jaw in La Naulette cave near Dinant, Belgium, in 1866, the hypothesis that this Neanderthal was simply an abnormal case was seriously called into question. The absence of a chin—considered an apelike feature at the time—on the La Naulette jaw was immediately striking. For the French anthropologist Ernest T. Hamy, the finds of La Naulette, Gibraltar (in 1848), and Neanderthal belonged to a single species, even a single race, which, he said, was the first of the fossil human races.

The site of La Ferrassie, in the Dordogne, was discovered towards the end of the last century. After ten years of methodical excavations, on 17 September 1909, Denis Peyrony uncovered the tibia and femur of the first of the eight Neanderthal skeletons buried in the great rock shelter some 40,000 years ago. Above: Two Neanderthal skulls.

After the discovery in 1886 at Spy, Belgium, of two skeletons—one of which was almost complete—that again displayed the same features, even the most hardened skeptics had to admit that the Neanderthals were not idiots or a hoax, and that another kind of human had indeed walked the earth before us.

The So-Called Cro-Magnon "Race"

In 1868 the world recognized the existence of another type of human with the discovery of five skeletons at the back of the rock shelter of Cro-Magnon, very close to the village of Les Eyzies in the Dordogne region of southwestern France. According to Louis Lartet (1840–99), son of the paleontologist Edouard Lartet, who was entrusted with the excavation of Cro-Magnon by the Ministry of Public Instruction, the position of the skeletons and the distribution of the remains accompanying them provided evidence that they had been deliberately buried. Six years after their discovery, in 1874, these people with anatomically modern traits were considered the prototypes of a new race, the so-called Cro-Magnon "race." Although they were associated with Paleolithic tools—the Paleolithic stretched from two million to 10,000 years ago—the antiquity of these Cro-Magnon people—who succeeded Neanderthals in Europe—was called into question, in particular by the French prehistorian Gabriel de Mortillet (1821–98). He could not bring himself to admit that Paleolithic people were already burying their dead. ("Paleolithic" and other terms can be found in the glossary, which starts on page 148.) Some French people who harbored ill feelings towards Germany after the Franco-

In 1868 workers discovered a rock shelter at Cro-Magnon, filled with bones and worked tools. Louis Lartet undertook the first excavations. Several adult human skeletons, which became the type specimens for the Cro-Magnon "race"—meaning that they were designated as having the defining characteristics of the group—were found, including the skull below, nicknamed the "Old Man."

The Cro-Magnon rock shelter.

The discovery of the Cro-Magnon people caught the imagination of illustrators of popular educational works, who portrayed what they supposed to be the customs of these "indigenous savages." Yet the Cro-Magnon people had many of the features of modern humans, as is shown in both this recent reconstruction (below left) and the skull of the "Old Man" of Cro-Magnon, with its vertical brow and protruding chin. Below, opposite, and inset opposite: The site of Les Eyzies in the Dordogne.

Prussian War of 1870–1 were delighted that a human type with a well-developed physiognomy and fine and elegant features had been discovered in their country, while the uncivilized Neanderthals, who still gave the impression of being ape-men, had been found in the country next door.

There still remained the task of discovering the makers of the oldest worked flints, collected most notably by Boucher de Perthes; no authentic vestiges of these toolmakers had yet been found.

Did the "Missing Link" Live in Asia?

Those who were convinced that humankind was descended from apes began searching for the missing link between the great apes and ourselves.

In 1868 Ernst Haeckel placed a Papuan at the end of his human family tree; he was convinced that the Papuans and Melanesians were closely related to the ancestral type, his hypothetical *Homo primigenius*. Struck by the resemblance between human embryos and those of gibbons, which live in Southeast Asia, Haeckel concluded that there had been "Gardens of Paradise" in this part of the world, on a continent that was partly submerged beneath the present-day Indian Ocean. This land would have corresponded with the "Lemuria" of the English scholar Philip Lytley Sclater (1829–1913).

Thomas Huxley, the passionate advocate of Darwin's ideas, believed that there was no distinction between humans and other animals, especially apes. In his book *Evidence as to Man's Place in Nature*, he demonstrated particularly that there are more differences between the hand and foot of a gorilla and an orangutan than there are between those of a gorilla and a human. Humans were simply a family within the order of primates.

Below, left to right: Skeletons of a gibbon, an orangutan, a chimpanzee, a gorilla, and a human.

Ernst Haeckel taught that our distant ancestors resembled gibbons. This photograph shows him holding the hand of a gibbon skeleton in a symbolic gesture. Standing behind it is the skeleton of a child.

In contrast to Haeckel, Thomas Huxley and Charles Darwin, respectively, put forward the hypothesis—in two remarkable books published a few years apart, *Evidence as to Man's Place in Nature* (1863) and *The Descent of Man* (1871)—that our primitive ancestors must have lived in Africa rather than Asia since chimpanzees and gorillas were, in their view, humankind's closest relatives. It was not long before finds in Asia—in Java—were made, and for a while they appeared to support Haeckel's point of view.

Anticipating the discoveries, Haeckel named the missing link—with his customary audacity—*Pithecanthropus alalus*, that is, silent ape-man.

Eugène Dubois Set Off on the Ape-Man's Trail

During this period a young Dutch anatomist by the name of Eugène Dubois (1858–1940), a keen reader and admirer of Haeckel's, enlisted in the Dutch army medical corps and in 1887 embarked for Sumatra and, later on, for Java, which were then Dutch colonies. Dubois was thoroughly convinced that humankind's

In reconstructing the genealogical tree of all living organisms, Haeckel relied on the embryogenic method. He adopted for his own ends the law stating that an animal's embryogenic development is merely a brief and rapid recapitulation of the successive stages of its evolution.

ancestor was descended from a primitive gibbon, and he was fired up by the idea of finding relics of these ape-people. The island of Sumatra, as Dubois was well aware, was indeed inhabited both by gibbons and orangutans. But only gibbons lived on Java.

It is astounding to think that, after mere months of relentless searching, Dubois would have been able to discover the remains of a hominid (a neutral term for a human being that does not specify time or gender) on Java, when one considers how extremely rare such finds are. It is true, however, that during these long months he amassed no fewer than 12,000 fossil bones. In 1891, having already found a jaw fragment and an isolated tooth, he unearthed a skullcap of vaguely human appearance as well as an isolated tooth at the south-central Javan

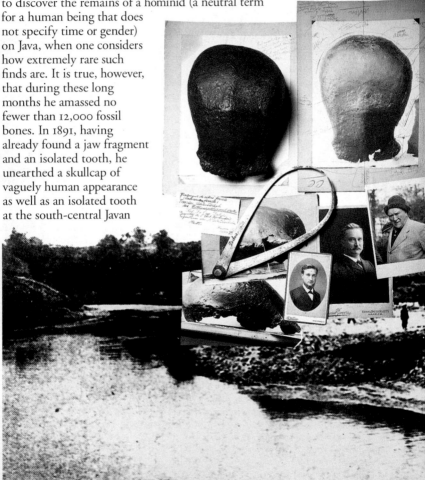

Below: The skullcap of *Pithecanthropus erectus*, Java Man, a drawing by Eugène Dubois, and photographs of the find and of Dubois himself at various ages.

Eugène Dubois had just arrived in Sumatra when he learned that a human skull, which turned out to belong to *Homo sapiens*, had been discovered near Wajak, Java (landscape left). He immediately set off for Java and soon discovered the human remains at Trinil on the banks of the Solo River (below) at the foot of Lawu-Kukusan volcano. Dubois was undeniably blessed with incredible luck, but he also had the good fortune to be helped by the fifty workers he had recruited for the Trinil dig alone. He wrote to Haeckel from Batavia to announce the happy news of the discovery of the missing link, which provided final confirmation of the zoologist's bold predictions. Presented at the third international congress of zoology in Leiden, Holland, in 1895, the remains of *Pithecanthropus erectus*, Java Man, were the subject of endless discussion for decades.

town of Trinil, under the alluviums of the Solo River.

The skull had a very low, receding brow and prominent brow ridges. To Dubois' astonishment, when viewed from above it exhibited marked narrowing behind the eye sockets. Curiously, he at first tried to establish a link between these remains and a chimpanzee. Since

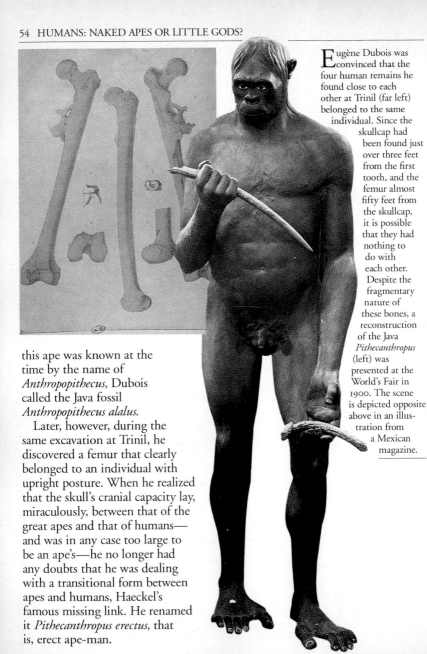

Eugène Dubois was convinced that the four human remains he found close to each other at Trinil (far left) belonged to the same individual. Since the skullcap had been found just over three feet from the first tooth, and the femur almost fifty feet from the skullcap, it is possible that they had nothing to do with each other. Despite the fragmentary nature of these bones, a reconstruction of the Java *Pithecanthropus* (left) was presented at the World's Fair in 1900. The scene is depicted opposite above in an illustration from a Mexican magazine.

this ape was known at the time by the name of *Anthropopithecus*, Dubois called the Java fossil *Anthropopithecus alalus*.

Later, however, during the same excavation at Trinil, he discovered a femur that clearly belonged to an individual with upright posture. When he realized that the skull's cranial capacity lay, miraculously, between that of the great apes and that of humans— and was in any case too large to be an ape's—he no longer had any doubts that he was dealing with a transitional form between apes and humans, Haeckel's famous missing link. He renamed it *Pithecanthropus erectus*, that is, erect ape-man.

The Discovery of Java Man Marked the Birth of Human Paleontology

Although the importance of the find was hailed on all sides, and people paid tribute to the man who had discovered it, skepticism remained widespread. For some, the Java fossil was a somewhat primitive native; for others, it was merely a large gibbon. In Virchow's view, the narrowing behind the eye sockets had never been found in humans: It was seen only in apes. However, Dubois continued to claim that the enigmatic form from Java was clearly a transitional stage between apes and humans. Discouraged and bitter about the fierce opposition to his theory, he permitted only a few rare visitors to examine his finds.

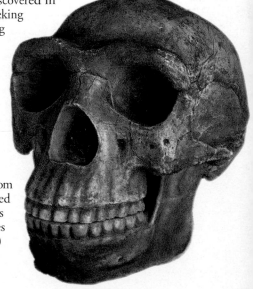

Dubois denied that there was any resemblance between the skulls found in Trinil and Peking (below), which he regarded as representing the Neanderthal race.

At the turn of the century, new expeditions went to Java, but they did not make any additional finds. The controversy dragged on interminably. Dubois grew weary of the struggle and of being misunderstood and finally refused anyone access to his fossils.

Meanwhile, a skull had been discovered in 1929 at Zhoukoudian cave near Peking (Beijing), China. Known as Peking Man, it displayed undeniable affinities with the skull from the Java *Pithecanthropus*. In the 1930s Ralph von Koenigswald, a German paleontologist (1902–82), brought to light more *Pithecanthropus* remains in Java, first not far from Trinil and then at Sangiran. New studies showed that the Peking and Java remains came not from an ape-man but from a true human, who was then named *Homo erectus*. Today the two types are regarded as different subspecies —*Homo erectus erectus* (from Java) and *Homo erectus pekinesis* (from China). Human paleontology had been born.

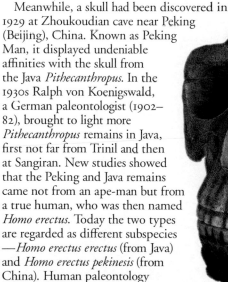

Are humans descended from apes? Naturally, people are not descended from present-day apes any more than we are descended from our cousins. But paleontology and all the disciplines of the biological sciences have taught us that humans and modern great apes had common ancestors several million years ago.

CHAPTER III

OUR ANCESTORS' ANCESTORS

This footprint (opposite), 3.8 million years old, proves that a bipedal upright hominid existed long before the first humans. There is evidence of a big toe alongside the other toes and the double curve of the arch can also be detected. Right: A chimpanzee.

The Protein Clock

In the early 1970s scientists studying modifications in the structure of blood proteins in primates calculated how long ago the lineage leading to the great apes (chimpanzees and gorillas) in the family Pongidae diverged from the human lineage. This "protein clock," as it would soon be called, was based on a constant rate of protein evolution during the history of the primates. When we measure temperature, the point chosen for zero is arbitrary and depends on whether one refers to the Kelvin, Celsius, Fahrenheit, or Réaumur scale.

The term *ape* covers a wide category of primates. The suborder anthropoids, which includes all monkeys, apes, and humans, contains all the descendants of a common ancestral type. Although opinions differ on the antiquity of the common ancestors, there is no doubt that they were apes.

Numerous attempts have been made to teach the great apes language. Today researchers study the abilities and mental processes brought into play by the use of a language based on meaningless symbols that the animal has to combine and reuse.

Similarly, in the case of the protein clock, it was decided to make zero the point when the Old World monkeys separated into the superfamilies hominoids (gibbons, gorillas, chimpanzees, and humans) and cercopithecoids (macaques, colobuses, baboons, and so on). At the time this date was estimated to be thirty million years ago. Working from this figure, it was calculated that the human line diverged from the African great apes about six million years ago.

Both humans and chimpanzees have twenty-three pairs of chromosomes. Thirteen of these are identical, but the other ten are organized differently, and on them some segments are completely reversed (diagram opposite).

Closer and Closer Cousins…

Although other dates were subsequently proposed, using data from genetics, cytogenetics, serology, and immunology, it was later confirmed that humans and great apes had diverged only recently; indeed, further confirmation was provided by their extreme genetic similarity—as we have seen, they share almost ninety-nine percent of their chromosomal material. This one percent difference was sufficient to produce significant differences in later generations, in terms of both anatomy and behavior, which finally resulted in the development of humankind.

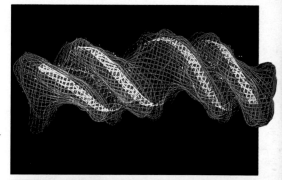

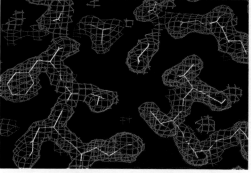

While the exact number of years that separate us from this event is still disputed —at present neither paleontologists nor molecular biologists agree about the actual date—the important point is that the great apes and humans are closely connected. After all, it has recently been noticed that chimpanzee communities share with us a mother-son incest taboo, a cerebral capacity for elementary

Much information can be gleaned from the study of chromosomes—made up of DNA molecules wound into a double helix (top)—and proteins (above).

language, a developed social organization, an intolerance between males, and even a consciousness of their own identity.

Geographical Distribution of Early Primates

The early history of humankind merges with the history of all the primates—tarsiers, lorises, lemurs, and, later, monkeys and apes. Sixty million years separate one of the last primates, humans, from the first known primate.

Although the oldest primate did not live at the same time as the last dinosaurs, it was part of a procession of archaic primates that bore far more resemblance to rodents than to small lemurs because of the peculiar shape of their incisors. At this time, the early Tertiary period (the Paleocene epoch, which began about sixty-five million years ago), they colonized Europe and North America, because there was then a land bridge linking arctic Canada, Greenland, and Europe, which thus formed a single continent. Barely five million years later, two new groups of prosimians lived on this continent. These were prosimians of a far more modern appearance: the

The origin of the lemurs (below and opposite), who live on Madagascar, is unknown because no fossil of the group has yet been found on the island or anywhere else. Lemurs constitute a very diverse group, with about thirty surviving species. Such diversification resulted from the fact that Madagascar separated from Africa a very long time ago, forming a real Noah's Ark where the animals could thrive without competition.

Certainly Madagascar split away more than thirty-five million years ago, since it is then that the first apes appeared in Africa. They never existed on the island.

tarsiiforms and the
adapiforms; the first remains
of these animals to be found
were discovered in 1822 in the gypsum quarries
of the hill of Montmartre. In the 19th century they
were considered to be the origin of the anthropoids,
the suborder that includes all the apes, monkeys, and
humans, but the question over the origin of the apes,
although debated, was not resolved.

As for the sudden appearance of these two groups in
Europe and North America, an enigmatic phenomenon
in itself, it too would have remained the subject of bitter
controversy if North Africa had not recently produced
fossils of prosimians that are older than those of the
northern continent, since they are more than sixty million
years old. This unexpected discovery constitutes a serious
argument in favor of an African origin for not only the
prosimians but also all the apes.

Whatever the validity of
these new hypotheses, almost
all the prosimians—both
adapiforms and tarsiiforms—
were to disappear suddenly
around thirty-five million years
ago. Their abrupt extinction was
certainly caused by a profound
deterioration in the climate.

The *Leptadapis* (skull
below) is one of the
best-known adapiform
primates. Like the rest of
its genus, which lived
forty million years ago, it
had enormous eye sockets
on the front of its skull,
which gave it stereoscopic
vision, required for
jumping and running
from branch to branch.

The Key Role of Africa and Arabia

From then on, the main theater of primate history was to shift, and events would take place further south, in the tropics of Africa and Arabia. It was here that the oldest known apes appeared, between thirty and forty million years ago. Whereas almost all African fossil apes come from the Fayum region, southwest of Cairo in Egypt, the remains of apes from the Arabian peninsula were discovered recently in the stony mountains of Dhofar, halfway between the Persian Gulf and the Red Sea in the sultanate of Oman. Distant ancestors of the great apes and hominids, they were the first to possess, like us, a set of thirty-two teeth, after the loss of a premolar in each half-jaw. Being tree

It was perhaps in a landscape like this—a humid tropical mangrove forest stretching along a seashore, the scene of occasional intense volcanic activity—that our most distant ancestors lived.

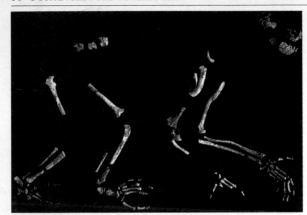

Working in the Great Rift Valley, paleontologists have discovered numerous remains of fossil apes, some of them very complete, like the famous *Proconsul africanus* skeleton (above left) found on Rusinga Island in Lake Victoria by Mary Leakey in 1948.

Skull of *Aegyptopithecus* (left), an ancestor of Old World anthropoids.

dwellers and good climbers, these quadrupeds, the size of small macaques, probably inhabited the tropical forests that bordered the rivers and deltas.

It is curious that no fossil apes have been found in the Old World from between twenty-two and thirty million years ago. After that period they suddenly reappear in East Africa.

Then, at this same time the Old World would be shaken by a series of movements in the earth's crust, including the upthrust of the Alpine chain, the rise of the Himalayas, the beginning of the enormous faults in the Great Rift Valley of East Africa, the opening of the Red Sea, and, most importantly, the collision of the two great continents of Asia and Africa, which was to have such a profound effect.

From Forest to Savanna

Around fifteen million years ago, major climatic changes on a global scale were to bring about the birth of a new type of landscape in East Africa: the savanna, into which many animals quickly moved. Leaving the relative protection of the forest environment that they had occupied until then, some apes would spread throughout the Old World, in Europe and Asia, taking advantage of the existence of a land bridge that

existed between Africa and Asia after the collision of these two continents.

In Asia the changes in habitat resulted in some apes evolving into orangutans several million years later, while in Africa they found expression in a new diet and, most notably, in locomotion.

In these savannas with few trees, tree-dwelling animals were eventually forced to descend. As they started to spend more and more time on the ground, they gradually adapted their mode of locomotion. Around five million years ago bipedalism became the primary form of locomotion in the australopithecines, the early primitive hominids. The reason for this major change in behavior was increasing severe drought to the east of the African Rift Valley.

As for the more conservative gorillas and chimpanzees, they would take refuge in the more humid environments of West Africa.

Around five million years ago huge mountains were formed in Africa, creating a barrier that dramatically reduced the amount of rainfall to the east. The result was these very open savanna landscapes (bottom) on the great plains of East Africa.

The Great Rift Valley, the enormous fault that cuts through East Africa for more than eighteen hundred miles, is clearly visible on this satellite image.

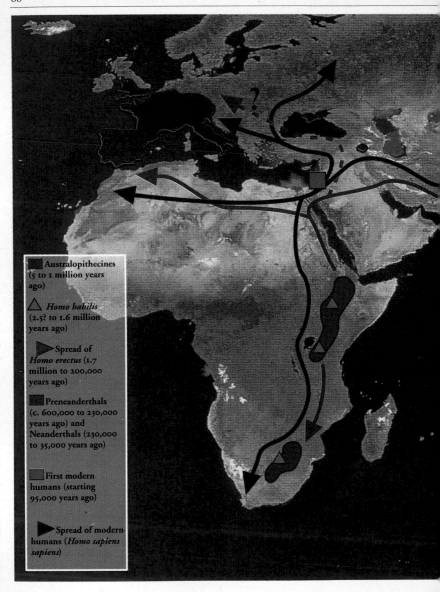

Australopithecines
(5 to 1 million years
ago)

△ *Homo habilis*
(2.5? to 1.6 million
years ago)

▶ Spread of
Homo erectus (1.7
million to 200,000
years ago)

Preneanderthals
(c. 600,000 to 230,000
years ago) and
Neanderthals (230,000
to 35,000 years ago)

First modern
humans (starting
95,000 years ago)

▶ Spread of modern
humans (*Homo sapiens
sapiens*)

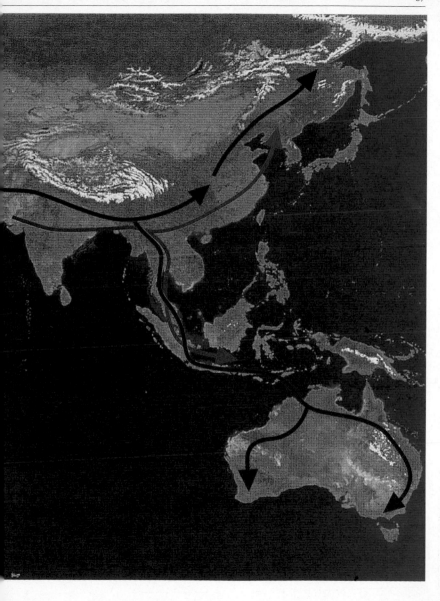

On 7 February 1925, Professor Raymond Dart, an Australian-born anatomist working in Johannesburg, published in the journal *Nature* the description of a fossilized skull that had been discovered at a limestone quarry at Taung, South Africa. He named the find *Australopithecus africanus* (southern ape from Africa), and in his view it was the missing link between apes and humans.

CHAPTER IV

FROM THE TAUNG CHILD TO LUCY

The reconstruction of an australopithecine couple who lived in Afar, Ethiopia, more than three million years ago (left) and the sign for the famous Olduvai Gorge in Tanzania (right) symbolize the adventure of Africa, the place where our most distant ancestors lie buried and where paleontologists try to piece together the puzzle of our evolution.

Australopithecus, **Africa, and Humans**

Dart's controversial conclusions were not accepted immediately. His missing-link theory proved to be a red herring, though he was correct in thinking that the remains belonged to the oldest-known human ancestor to be discovered in Africa. This continent, which until then had been silent on the problem of human origins, would become the scene of the most avid paleontological excitement: In fact, all of the earliest-known hominids, known as australopithecines, have turned up in Africa. As Charles Darwin had already said in *The Descent of Man*, this continent was undoubtedly the place where our first ancestors emerged.

Known today to have existed five million years ago—perhaps even longer ago—in East Africa, the australopithecines were small bipedal creatures that became extinct less than a million years ago, if the age (less than 900,000 years) of the deposits at Taung Cave in South Africa is confirmed.

In 1924, during a mining operation at a limestone quarry at Taung, South Africa, the fossil skull of a child was blasted out of the rock from a depth of about fifty feet. After being used for a while as a paperweight by the mine's director, it finally came into the hands of Raymond Dart (left). In February of the following year he announced the discovery of an extinct ape intermediate between living anthropoids and humans, which he named *Australopithecus africanus* (reconstruction below) and claimed was the first species of a new family of primates—the *Homo-simiadae*.

Discovered at the beginning of this century, the sites at Omo, Ethiopia (left), were explored in 1932 by the French paleontologist Camille Arambourg. In 1967 an international expedition went to Omo and collected almost 100 tons of fossil bones.

The Eldorado of the East African Rift

Some thirty years later, it was the turn of the Great Rift Valley, the series of enormous faults running from the Middle East down through Africa. The East African section would witness a tremendous deployment of equipment and researchers. The Rift Valley, which was formed almost twenty million years ago, is marked by a string of great lakes. Over time, many thousands of feet of sediments have accumulated in it, making this area an ideal trap for fossilization. This exceptional phenomenon accounts for the fact that all the ancient sites containing fossil hominids in Africa—except for a few sites in South Africa—are located along this series of faults.

As the years have passed and the earth's crust has shifted, these sedimentary layers, which were at first piled on top of one another, have often toppled over until they began to look like the pages of an open book. The finest

This view of the Great Rift Valley shows the immensity of the scar that cuts through East Africa from north to south. It was caused by the slow separation—at the rate of a few hundred feet per million years—of the African plate to the west and the Somalian plate to the east.

example of "open pages" is provided by the famous sites in Ethiopia's Omo River basin. Here the fossil remains of our ancestors and their numerous worked tools can often be found lying on the ground. Volcanic deposits interspersed with these geological layers make it possible, through the radioactive elements they contain, to date this evidence of our remote history.

A group of australopithecines with the carcass of an antelope, which they must defend from hyenas (below).

Gracile and Robust Australopithecines

All these sites in Africa have yielded two distinct groups of australopithecines. One is comprised of three small "gracile" forms: the smallest, *Australopithecus ramidus*, which was the most apelike hominid ancestor, dating to around 4.4 million years ago, *Australopithecus afarensis*—best represented by the well-known find called Lucy (see page 84)—and finally *Australopithecus africanus*. The other group is comprised of three "robust" forms (*Australopithecus aethiopicus, Australopithecus robustus,* and *Australopithecus boisei*).

They all had a concave and forward-jutting face, a poorly developed brow, a more or less pronounced ridge above the eyes, and molars and premolars that were enormous (except in the case of *Australopithecus ramidus*). None had a chin. Though they were equipped with a brain that was still small (450 to 550 cubic centimeters, approximately 27 to 34 cubic inches) compared with ours (1400 cubic centimeters, about 85 cubic inches, on average), these hominids were already moving around upright on their two legs. We know, however, that they had not lost their ability to climb trees.

The robust australopithecines, who were distinguished from their gracile counterparts by the presence of a very marked crest on the top of their skull, were strange beings with enormous cheek teeth, operated by a powerful jaw. They had a cranial capacity of 550 cubic centimeters, or about 34 cubic inches. Strictly vegetarian, living on roots, nuts, and seeds, they disappeared, with no descendants, around one million years ago.

The gracile australopithecines, some of which left

Discovered in 1911 by a German entomologist, Olduvai Gorge (opposite above) is one of the centerpieces of African prehistory. From the 1930s the famous British scholars Louis Leakey (1903–72) and his wife Mary (1913–) devoted themselves to exploring the site, where there were outcrops of major lacustrine deposits (from lakes) dating to the last two million years. Their extensive work in the area earned them great distinction. In 17 July 1959 Mary Leakey discovered one of the most famous australopithecine skulls (opposite left), which Louis called *Zinjanthropus boisei*—Zinj being an ancient East African name. It was a robust australopithecine, the first to be found in this part of Africa.

the forest environments along the rivers and gradually reached the savannas, are considered by some to be the makers of the first worked tools. Hunters of small game, and occasional meat eaters, they were fairly omnivorous. Contemporaries of the robust australopithecines, these far weaker beings—ranging in stature from about three to four feet and in weight from sixty-five to ninety-five pounds—had a slightly smaller brain.

In the Wake of *Homo Habilis*

While some paleontologists regard the australopithecines as an evolutionary cul-de-sac, in the sense that they are not considered to be direct ancestors of humankind, others believe that they may have given rise to the first humans, who differed from the australopithecines in several respects—their larger size, a flattened face with no ridge above the eyes, smaller molars and premolars, and, most importantly, a much bigger brain, with a capacity up to 800 cubic centimeters, or about 50 cubic inches. In 1964 Louis Leakey named this new type of hominid *Homo habilis*—skillful person—because the study of its fingers revealed an ability to grip that was sufficient for working stone tools.

The fossilized remains of this new species, which came from Olduvai Gorge in Tanzania, were dated to 1.8 million years ago. Leakey's announcement immediately provoked some very passionate criticism concerning the attribution of these bones to a human species, and much discussion of the inevitably arbitrary definition of humanity ensued. In order to solve the problem, scientists enumerated one hundred anatomical features peculiar to the human genus. This list of characteristics, drawn up especially for paleontologists—who never have more than a selection of disparate fragments of fossils at their disposal—left out, with good reason, all other biological features, not to mention behavioral and cultural traits.

In 1961 Mary Leakey discovered in Olduvai Gorge a skull of a new human species, *Homo habilis*, the oldest representative of the genus *Homo*. Eleven years later her son Richard would in turn reveal a far more complete skull of *Homo habilis*, the famous ER-1470, best known by its catalogue number at the National Museum of Kenya. The head has been reconstructed (below) by adding the missing parts and the tissues and muscles of the face. The imprint of the brain, whose size is up to forty-five percent greater than that of the australopithecines, suggests that it might have possessed the neurological bases of speech. In any case, the discovery of contemporaneous circular structures provides evidence that *Homo habilis* had a relatively complex culture.

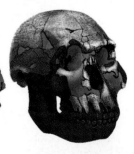

Homo habilis (left) undeniably had an extensive tool kit, mostly made of stones worked on one or both faces. The technique of striking flakes off a stone is known as the Oldowan industry, after the site of Olduvai, where the first such tools were found.

The ER-1470 skull, which was found at Koobi Fora near Lake Turkana, is made up of hundreds of bone fragments; fitting them together took several weeks of patient, skilled work. Below is a reconstruction of Homo habilis.

The Significance of Tools in Defining Humans

In the course of time, as the great East African sites were explored, a very ancient stone tool–making tradition was discovered. It consisted of small stones, or pebbles, that had barely been retouched and of broken quartz flakes. Dating back almost three million years and probably made by the australopithecines, the earliest tools used in the world provide irrefutable evidence of the development of thought—with all its cultural and even social implications— in beings that were not yet human.

Faced with the often contradictory interpretations of fossil hominid remains, the prehistorian considers that a tool—that is, a deliberately modified natural

object—speaks volumes about the degree of "hominization" (or the process of becoming human) in the person who made it.

However, it is extremely hard to identify exactly who this person was because, in the case of East Africa, we are more or less certain that two, or even three, types of hominids coexisted during this crucial period. And this invention of tools was made independently by at least two of them. In fact, we can no longer be sure that the use of tools is a factor in what makes us human. For this reason many paleontologists continue to define humans in terms of biology rather than cultural behavior. However, it is certain that the making of tools, even very primitive ones, requires a logical series of actions conceived by reflective thought.

While it is still very difficult to know whether the oldest hominids altered or retouched organic materials such as wood, bones, or teeth, it is certain that the first objects to be made of stone date back almost three million years. Several sites in East Africa lay claim to this honor at present. Two of them are located in Ethiopia —at Afar and Omo.

At Afar, worked stones (that is, stones from which flakes have been struck off with another stone), cores (which are quite similar), and flakes have been found buried in volcanic ash dating to more than

The oldest stone tools known in the world come from three sites: Kada Gona in the Afar Triangle of Ethiopia; localities 71 and 123 at Omo; and Koobi Fora to the east of Lake Turkana in Kenya. Comprised of worked stones, in the case of Kada Gona (below), or small flakes of quartz as at Omo (left), these still rudimentary tools are dated to between 2.1 and 2.6 million years ago. In this photograph taken in the 1960s (opposite) Louis Leakey points out the very spot where, in 1929, he discovered countless tools of volcanic glass at Kariandusi in Kenya. After that Leakey would become increasingly convinced that he would find similar remains in Olduvai Gorge.

2.5 million years ago. The task of working a pebble was not performed in a haphazard manner; flakes were taken from one or both faces, and sometimes they even broke through the width of the stone. These tools, which were still very crude, were no doubt used for cutting hides, scraping bark and skins, or crushing bones and shells. Cores were stone blocks, generally of volcanic rock, from which flakes had been detached. Although the technique was identical to that of working stones, the goal here was to obtain the flakes themselves, which were extremely sharp when fresh.

At Omo, many little flakes, obtained from pebbles or from quartz nodules, have been found. While for a long time it was thought that the upright stance in hominids freed the hands for making tools, it now seems increasingly clear that two to three million years, or even more, separate bipedalism (and consequently the freeing of hands) from the first worked tools. It therefore appears that the appropriate mental faculties had to be developed before stone tools could be fashioned.

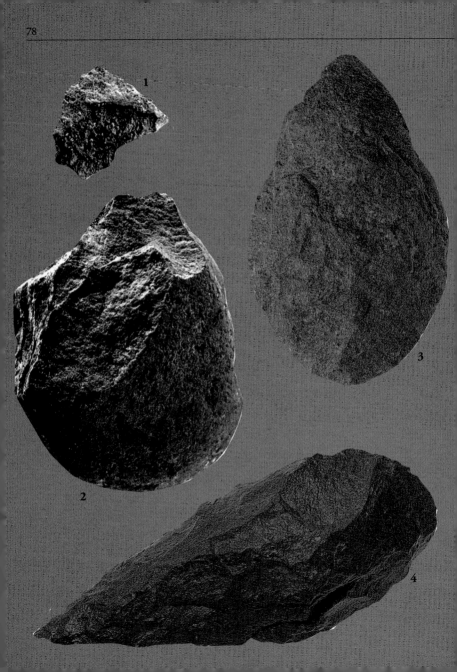

From Worked Pebbles to Harpoon Points

The typical tool of 1.9 million years ago was a stone that had been worked without special preparation, generally flaked on only one side (2) and often accompanied by crude flakes (1). These sorts of tools, found at Olduvai and made by *Homo habilis*, defined that rudimentary industry known as the Oldowan. Later on, pebbles were worked more thoroughly —some over their whole surface, resulting in polyhedrons or spheroids. Knapping techniques, which would gradually be organized around a function, evolved particularly slowly for several hundred millennia before humans— *Homo erectus* this time— managed to produce new tools in a wide variety of rocks (3, 4, 5). At first these were rather crude knives with a sinuous cutting edge, then, because of increasingly precise work, they became beautiful almond-shaped flints with a regular cutting edge (8).

The Invention of Tools

The Acheulean bifaces or handaxes (pages 78–9: 3 and 4) and cleavers, tools with a transverse cutting edge (5), were used for a very long time before other cultures started to employ tools made on flakes once again. These later cultures applied new pressure techniques (6, 7, 10, 11, 12), consisting of detaching flakes from the surface of a core (9) with a pointed instrument. The fabrication of such tools sometimes required a long series of preparatory stages. It was only towards the end of the last glacial period that the nomadic hunters of the Upper Paleolithic would learn to make tools that were increasingly sharp and pointed. With these tools bone and deer antler were worked to produce awls, eyed needles, harpoons, or spear points (14, 18).

13

16   17

18

14

15

Bipedal Locomotion

The increase in the size of the brain, the center of thought, was one of the main factors in making humans unique. In other words, it made them different from other animals. Indeed, some scientists have even chosen the arbitrary figures of 700 to 800 cubic centimeters (about 43 to 50 cubic inches) to delineate the point at which apes become human. To attain a cranial capacity higher than this means crossing the "cerebral Rubicon." Of course, in a more or less continuous evolutionary process, such a cutoff point has no meaning. Nevertheless, a number of scholars have always thought, albeit without proof, that an upright stance and bipedalism preceded the increase in the brain's size.

Sophisticated forms of imaging, like these broken-down images of walking (below) employed in studies of the biomechanics of fossil human locomotion or scans that permit a researcher to see inside a skull (bottom), demonstrate the specialized nature of paleoanthropologists' work.

When, after the First World War, the skull of the Taung Child was presented to Professor Dart, he was convinced that the young individual to which it belonged usually stood upright. In fact, this theory was not based on any evidence, because the appropriate bones—the limb bones and the pelvis—which would have provided conclusive proof, were missing.

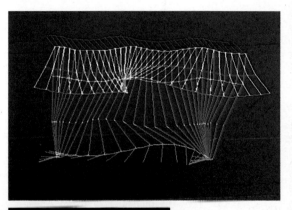

Support for Dart was to come from a Scottish physician and part-time paleontologist, Robert Broom (1866–1951), who, a few years after the Second World War, discovered several pelvic bones as well as a spinal column in two other South African sites—

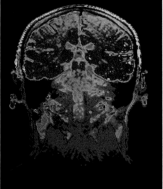

As the genus *Australopithecus* had been founded on the skull of a child (who was, therefore, not fully developed), further fossil evidence was needed to confirm Dart's hypothesis. In 1936, at Sterkfontein, near Johannesburg, Dr. Robert Broom (opposite) became the first to find the skull of an adult australopithecine.

Sterkfontein and Swartkrans. These finds provided decisive evidence for the existence of an upright stance in australopithecines.

The resounding confirmation of bipedalism in remote prehistory came from the 1974 discovery by Donald Johanson and his team at Hadar, Ethiopia, of an astonishingly complete skeleton, known as Lucy.

Two years later humanlike footprints were found at Laetoli in Tanzania in volcanic ash that has been dated to 3.6 million years. None of the paleontologists working at Laetoli for two seasons had even suspected their existence. Then, on a September evening in 1976 one of them, British scientist Andrew Hill, noticed two prints that resembled those made by human feet among numerous tracks of birds, elephants, rhinoceros, and other mammals.

In 1978 Mary Leakey uncovered a series of prints with two parallel trails. Analysis of the footprints showed that they had been made by two individuals, perhaps walking side by side, one of whom was 4 feet 6 inches tall, while the other was smaller and measured just under 4 feet. Thanks to the exceptional state of preservation of the footprints, it is possible to prove that these individuals were walking perfectly upright. It is clear from this evidence that bipedalism is an extremely ancient attribute and probably goes back well beyond five million years.

Below: Paleo-anthropologists Yves Coppens and Donald Johanson (right).

In the Afar Triangle of Ethiopia Donald Johanson's Franco-American team (led by Maurice Taïeb) unearthed fifty-two bone fragments of an australopithecine they named Lucy, after the Beatles song "Lucy in the Sky with Diamonds," which was popular at the time. Conclusive proof of bipedalism came from the fragile remains of this young female representative of *Australopithecus afarensis*; forty percent of her skeleton (opposite above) has been found. Additional support came from these footprints (opposite far left) discovered at Laetoli, Tanzania, which were left by members of Lucy's genus a few hundred thousand years earlier.

Left: A reconstruction of Lucy.

While *Homo habilis* finds proved that the genus *Homo* had originated in Africa, several human fossils generally attributed to *Homo erectus* testify to its appearance there 1.7 million years ago. Setting out from the sub-Saharan regions where this hominid lived for several hundred thousand years, *Homo erectus* eventually inhabited North Africa, Asia, and probably Europe.

CHAPTER V

THE LONG MARCH OF *HOMO ERECTUS*

A rational explanation for the origin of fire was given in the 1st century BC by Roman poet Lucretius: "It is lightning that brought the first flame down to earth for the use of mortals." In the picture opposite *Homo erectus* discovers fire. Right: Bust of Peking Man.

Homo Erectus: A Close Relative

Homo erectus, those humans who are sometimes called pithecanthropes—a name that should really only be used for the Java *Homo erectus*—gradually spread throughout the temperate regions of the Old World. The oldest *Homo erectus* populations were certainly descendants of *Homo habilis* and showed features of a more evolved species. They appeared approximately 1.7 million years ago in the Lake Turkana region of Kenya.

Outside Africa, some would reach Asia—Java, to be exact—around one million years ago. However, there is some evidence suggesting that they may have arrived in Asia and even Europe many years earlier. Recent research on the child's skull of Modjokerto and the human fragments from Sangiran discovered in Java in 1936 and 1974, respectively, points to ages as great as 1.8 and 1.6 million years.

Although they occurred in a vast geographical area, these hominids all shared the same physical features, which suggests that they belonged to a single species. However, some anthropologists believe that *Homo erectus* did not actually exist as a biological species but has merely been conceived to explain human evolution. The debate continues.

As far as physical traits went, the stature of *Homo erectus* was close to our own—some individuals reached a height of five and a half feet. However, their cranial capacity, which varied between 775 and 1250 cubic centimeters (about 47 to 76 cubic inches), was much smaller than ours today. While they had robust jaws, and teeth that were still enormous, their skull presented a number of peculiar bony superstructures, such as a keel on its vault, and a thick brow ridge above the eye sockets, forming a kind of visor.

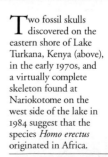

Two fossil skulls discovered on the eastern shore of Lake Turkana, Kenya (above), in the early 1970s, and a virtually complete skeleton found at Nariokotome on the west side of the lake in 1984 suggest that the species *Homo erectus* originated in Africa.

Occupying the stage of the entire Old World for almost 1.5 million years, *Homo erectus* mastered fire. They gradually became skillful hunters and at the same time set up the first habitations. They also developed a more sophisticated tool—the biface, a core worked on both sides. The regular cutting edge of this tool was used for a variety of different tasks, such as scraping, digging, and cutting. This industry formed part of the Acheulean tradition, named after the site where it was first found in St. Acheul, France.

Conquering Their Fear: The Control of Fire

The first undeniable traces of the use of fire occurred around 450,000 years ago. However, the origin of fire making may date back even further, to almost 1.5 million years ago, as shown

In 1966, during excavations carried out by French prehistorian Henry de Lumley near an alley named Terra Amata in Nice, several prehistoric levels of occupation were unearthed. Each layer corresponded to different seasonal camps made by hunters 380,000 years ago (above). In 1971 de Lumley discovered at Arago Cave near Tautavel (in the eastern Pyrenees) the oldest human skull in Europe. It dated back some 450,000 years (left).

by the site of Swartkrans. Several anthropologists stress the fact that humans first had to overcome their fear of fire, which they most certainly shared with all animals. The control of fire involved a major change in behavior, and the capacity to create and use it must have developed very rapidly. While we are relatively sure that *Homo erectus* repeatedly used embers produced by natural forest fires, we do not know how humans learned to make and maintain fire. Simple techniques, still in use today, using flint lighters or rubbing two sticks together, were undoubtedly discovered quickly by *Homo erectus*.

The hearth structures discovered on the hut floors at Terra Amata in Nice (left) constitute, along with those at Vértesszöllös near Budapest in Hungary and at Torre in Pietra, Italy, the oldest evidence for the control of fire. At Terra Amata the prehistoric humans set up their fireplace in the very center of the hut, in a small hollow dug out of the sand and protected from the prevailing winds by a little wall of pebbles.

Fire Brought People Together: The Beginning of Life in a Community

Although fire making in itself was not a basic need, it did help people control their world. The use of fire had many consequences. First, fire provided warmth—which actually may not have been very useful in this period. Second, as a source of light, fire enabled people to escape the rhythms of nature and, for example, to occupy the depths of caves. Third, it was also a powerful tool to be used in protecting themselves from wild animals. Finally, fire enabled food to be cooked, a phenomenon for which we have evidence from this period. Cooking implies that meals were eaten communally. Coming together to eat was an act of strong social or family

cohesion, just as it is today. It is clear, therefore, that the control of fire played a major role in the evolution of human societies.

It does seem that *Homo erectus* was a peaceful hunter-gatherer, as not one fossil bone from this long period bears any trace of violence. The first evidence of violence appears only during the Neolithic period, around eight thousand years ago, when cultures would become increasingly settled.

How Did Early Humans Communicate? When Did They Learn to Speak?

Thinkers and philosophers have been passionately interested in the subject from the earliest times, long before prehistorians ever looked into the puzzle. While some wondered if thought could exist without speech, those who commented on the Bible affirmed that Adam and Eve spoke Hebrew. Even though we do not have

Evidence of the first *Homo erectus* in China was discovered near Peking (Beijing) in Zhoukoudian cave, which was occupied for more than 200,000 years. After its roof collapsed, these hominids had to find shelter in the western part of the cave, which they then were forced to abandon after it filled up with sediment. They clearly knew how to use fire, as is shown by the existence of four ash layers, one of which is almost twenty feet thick in places.

all the answers to the problem, quite a few prehistorians are now convinced that *Homo erectus* communicated by speech with some degree of skill. The culture of these tool-makers who knew how to use fire and lived from hunting and gathering was certainly so complex that they must necessarily have given names to objects, plants, and animals, identified places, and exchanged ideas.

The extent of their activities undoubtedly called for a highly developed means of communication that was far more developed than that used by apes, though some people take issue with this point of view. Indeed, reconstructions of the vocal apparatus of a few fossil humans, albeit highly controversial, seem to indicate that the acquisition of language was a more recent phenomenon.

Be that as it may, even if this earliest language was rudimentary, comprising simple sentences pronounced slowly, it may have been accompanied by a large number of other acoustic or visual signals, such as cries, whistles, and facial expressions. These humans would also have been able to employ sign language, like that used by deaf people, or in some cases by New Guinea tribes or by Bushmen.

In humans, the larynx (above right, in red) is located lower in the throat than in the chimpanzee (above left). Above the larynx, the pharynx (in pink), which plays the role of an organ pipe, amplifies the sounds emitted by the vocal cords. In the course of human evolution, it was not until the larynx dropped, which in turn led to an increase in volume in the pharynx, that language could be acquired.

"Dragon bones," which were ground up and sold in the form of a powder known for its medicinal properties, have been used by Chinese apothecaries, it is said, since the Song dynasty (AD 960–1279). They are, in fact, the teeth of fossil animals.

The Misadventures of Peking Man

The first clues to the existence of fossil humans in China go back to the beginning of this century. At this time Chinese pharmacies sold individual teeth called "dragon bones," which, when ground into powder, were meant to treat practically anything. Then one of the teeth that came from the limestone quarries in the caves of Dragon Bone Hill, not far from what was then Peking (Beijing), turned out to belong to a primitive human. Methodical excavations were undertaken after the First World War in the biggest of these caves, and two teeth were found in 1926, followed by a third tooth the next year. Davidson Black, a Canadian professor of anatomy at Peking Medical

After three years' hard work at Zhoukoudian, in 1929, W. C. Pei, the leader of the Chinese team, found the first almost-complete skull of *Sinanthropus* (left); it was taken for analysis to the anatomist Davidson Black (below, with a pipe). On his left is G. B. Barbour, a renowned American geologist; on his right is Father Teilhard de Chardin, a scientific adviser to the Geological Survey of China.

College, studied the teeth and named the remains of this primitive human *Sinanthropus pekinensis*, that is, the Chinese man from Peking. Today it is known as *Homo erectus pekinensis*.

This discovery marked the start of feverish activity at the site. Major excavations lasted until 1937, when the political situation in China, at war with Japan, reached a crisis point. Just before the war, it was estimated that the remains of forty-five individuals had been unearthed.

In 1941, as Japanese troops drew threateningly close, China decided to send the remains of Peking Man to the United States. They were carefully packed and handed over to a detachment of American Marines who were to embark for the United States a few days later. But the warship on which the American soldiers are said to have embarked was sunk. It is now generally believed that the *Sinanthropus* remains are lying somewhere in the Yellow Sea, at a depth of less than 650 feet, along with the skeletons of the Marines.

Found in 1906 in a sand quarry near the village of Mauer, near Heidelberg (photograph, left), the massive jaw (below) of a hominid with enormous teeth lacks the protruding chin that characterizes modern humans.

The Traces of the First People in Europe

The oldest human fossil in Europe, a lower jaw found at Mauer, not far from Heidelberg, is no more than 600,000 years old. Not so long ago it was thought that Europe was colonized from Africa much later than Asia. However, in the last few years, several discoveries, limited at present to a few worked pebbles and flake tools, have led us to reconsider this view.

For the time being, the site of Chilhac has provided the oldest evidence of a human presence in Europe. Located in the Auvergne, Chilhac has produced several stone tools found with fauna dating back almost 1.5 million years. A great number of sites, both open-air and caves, dating from about 900,000 years, are known in France (Blassac, Soleilhac, Wimmereux, Vallonnet cave) as well as in other western and central European countries.

Nevertheless, it appears that Europe was rather sparsely populated during the long glacial period known as Mindel (450,000/400,000–300,000). Thanks to the milder climatic conditions that followed, during the 100,000-year-long Mindel-Riss interglacial, human settlement gradually became more dense. All the same, while it seems certain that Europe was populated at an early period, we still have much to learn about the hominids of the time.

Overlooking the plain of Tautavel, Arago cave made an ideal hunting camp for prehistoric people 450,000 years ago. On the steep slopes they hunted ancient forms of wild sheep, goats, and antelopes. In the valley below the cave and on the plateau above, there were grasslands where herds of aurochs and other early oxen and bison grazed.

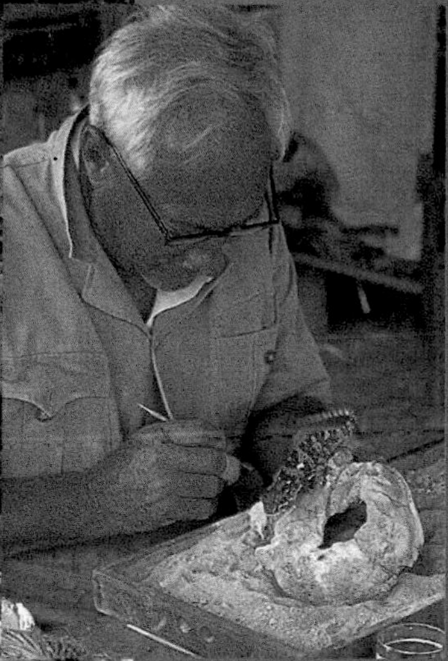

Excavating "Zinj"

Louis Leakey and his wife, Mary, pictured here with their son Philip, began their studies at Olduvai Gorge in the 1930s. In 1959 Mary Leakey discovered the *Zinjanthropus* skull in Bed I, the oldest at Olduvai. The excavations, carried out over an area of about 400 square yards, revealed an ancient occupation level. The bone fragments and worked tools abandoned by a group of hominids who had camped there more than 1.75 million years ago were rapidly buried beneath the clay left when the waters of the Olduvai lake rose.

Painstaking Work at Olduvai Gorge

Once they found it, the excavation of the *Zinjanthropus* skull at Olduvai was a long and exacting task. For nineteen days, in overpowering heat, Mary and Louis Leakey worked intensely, fired by emotion, to collect the four hundred bone fragments that remained of the skull. After the excavation, Louis Leakey (opposite) pointed out the exact stratigraphic position of the "Zinj" skull. The vertically dug walls show lacustrine deposits alternating with layers of volcanic ash.

The sites at Olduvai, like those at Lake Turkana (overleaf), yielded many bones of a strange elephant, the *Deinotherium*, which had no tusks in its upper jaw, unlike modern elephants. Curiously, the lower jaw curved down and was continued by two backward-curving tusks (left).

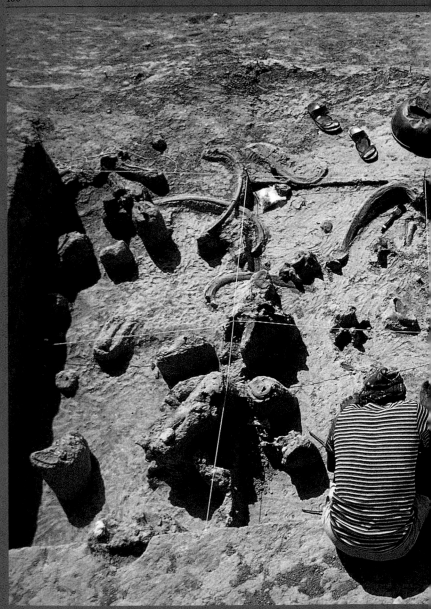

A Horizontal and Vertical Reading

Numerous prehistoric sites have yielded a high concentration of different remains, like this extraordinary tangle of mastodon bones (left), laid down in a random fashion nearly five million years ago. In cases like this it is essential to carry out meticulous excavations; every object is located on a horizontal plan using a grid of meter squares before being removed and catalogued (see pages 100–1). Detailed plans are used as the basis for analyzing and interpreting the associations between objects. The way in which bone fragments, like those of this rhinoceros (opposite below) found in Saudi Arabia, are arranged provides evidence about the direction and strength of the currents when they were buried. Pieces discovered are recorded carefully to give an overview of the site both horizontally and vertically through the superimposition of the different plans. In this way long-ago events are placed in a chronological framework.

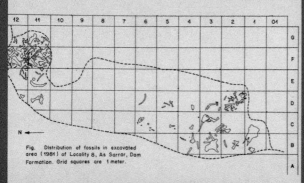

Fig. Distribution of fossils in excavated area (1981) of Locality 8, As Sarrar, Dam Formation. Grid squares are 1 meter.

In the last 200,000 years, major climatic changes have taken place, and the course of human evolution has become more complex: Technical innovations occurred, spiritual concerns emerged, societies became increasingly ritualized, and aesthetic feelings began to appear. Cultural evolution took precedence over biological evolution.

CHAPTER VI
FROM NEANDERTHAL TO LASCAUX

Some 15,000 years separate these two figurative paintings. The rhinoceros on the right symbolizes the interests of the hunters during the last Ice Age (about 110,000–10,000). The genre scene opposite bears witness to the scientific and didactic spirit that was emerging in the 19th century, an age that tried to make sense of the older, vanished world.

Among the artists who have depicted the life of prehistoric people, the Czech Zdenek Burian (1905–81) is preeminent. Though he was the heir to the academic painters of the late 19th century in his realistic and picturesque reconstructions, Burian also distanced himself from them in his more scientific and more humane vision, as in this depiction of Mauer Man (left).

A Difficult Transition: The Frustrating Paradox of Our Recent Ancestors

When, how, and where did *Homo erectus* give rise to *Homo sapiens*? The origin of *sapiens* is one of the hardest problems to solve in the whole of human paleontology. The period between 600,000 and 200,000, during which Europe was twice covered by an ice sheet (the Mindel glaciation and Riss glaciation), is certainly one of the most obscure periods in the history of humanity. It is somewhat paradoxical that whereas our earliest ancestors, the australopithecines, are relatively well known—more than a thousand specimens and dozens of skulls have been found—the human remains from this later period are rather rare. Moreover, doubts have been expressed about the dates attributed to these finds. Analysis of any fossil generally gives an approximate age based, in Europe, on the alternation of glaciations and milder periods, the interglacials.

Although most human remains have been found in western Europe and North Africa, it would be wrong to conclude that the early humans did not exist elsewhere.

Not only do the fossil beds of immense areas of the globe still remain unexplored, but the age of some sites is still uncertain (such as Broken Hill in Zambia and Hopefield in South Africa). Many sites of this period, which range in area from northwestern Africa to China, reveal, in their stone tool industries, the existence of these humans, the last *Homo erectus* populations.

In Europe, a dozen sites containing human fossils remain problematic for prehistorians. They include Bilzingsleben, Petralona, Mauer, Atapuerca, Boxgrove, Montmaurin, Vértesszöllös, Swanscombe, Steinheim, and Arago. What do they tell us?

Two Human Populations, Two Totally Different Lineages

Because of the incomplete nature of these remains, paleoanthropologists still disagree about the different stages in human evolution. Until recently it was thought that we were dealing with two populations evolving in parallel in Europe, one of them leading to the well-known Neanderthals, the other resulting in modern *Homo sapiens.*

The Neanderthals acquired a degree of technological complexity that is reflected in the evolution of their tool kit. Increasingly sophisticated tools were used to perform various functions —cutting, piercing, scraping. At the same time social organization became more developed, as each new technique had to be passed on to the group. Tools then became "social objects."

This quick drawing by the famous prehistorian Abbé Henri Breuil (1877–1961) summarizes in naive fashion his view of the social organization of Neanderthals.

A great number of female statuettes known as Venuses, sculpted in stone or ivory, and sometimes engraved (like the Venus with a horn from Laussel, left), have been found in Upper Paleolithic sites all over Europe and as far as Siberia. In Europe these Venuses often have exaggerated features— they are very fleshy, with marked steatopygia, enormous breasts, protruding stomachs, and broad hips. Facial features, as in the figurine opposite above right found in the last century in a cave at Brassempouy in the Landes (south-western France), are rare and predate by a few millennia the great period of cave art represented by the paintings at Lascaux. Without a doubt these statuettes show how women were viewed in the Upper Paleolithic culture: as matriarchs or as women-mothers with a magical power linked to the fertility that ensured the group's survival.

The term "modern" *Homo sapiens* is used because the Neanderthals were also the species *Homo sapiens* but were a subspecies; they were given the scientific name of *Homo sapiens neanderthalensis*. Recent studies have shown that only the lineage leading to the Neanderthals was actually present in western Europe. The Neanderthals, whose peculiar anatomical characteristics gradually appeared around 230,000 years ago within European populations, colonized all of peninsular Europe around 70,000 and subsequently reached the Near East and Central Asia. They disappeared about 35,000 years ago.

The other lineage, leading to modern humans, has to be sought outside Europe, perhaps in Africa. According to this theory, anatomically modern *Homo sapiens* originated in Africa, reaching North Africa and the Near East around 100,000 years ago. And it is precisely in this region of the world, in Qafzeh cave, near Nazareth in Galilee, that the oldest skeletons of modern humans have been found.

This double grave (opposite) of a young woman with a six-year-old child at her feet dates back almost 100,000 years. It was discovered in Qafzeh cave, in Israel, and is the oldest known burial. The excavations carried out at Qafzeh in 1933 revealed five human skeletons. Between 1965 and 1975 the French prehistorian Bernard Vandermeersch discovered more human bones, this time of six adults and seven children, many of whom—in particular a youth—had been deliberately buried. They displayed many similarities with the European Cro-Magnons, who appeared some 60,000 years later.

The techniques of reconstructing Neanderthals have been considerably improved (left) since the first portrait drawn by Schaaffhausen in 1888. Nevertheless, we have still not formed a clear image of many details of their physical appearance.

The use of natural objects from 40,000 years ago provides evidence for the first aesthetic or religious focus. The first pieces of jewelry, pendants often made of animal teeth pierced with a circular hole or marked with a groove so that they could be hung, appear 5000 years later, with the start of the Châtelperronian culture. The choice of these teeth, often of carnivores—foxes, bears, wolves, hyenas—suggests that the artisans hoped to benefit from the qualities

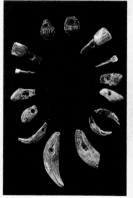

that they thought resided in these animals. Moreover, wearing such ornamental objects was perhaps intended to denote differences in age or sex, or even in an individual's social status.

The Time of the Würm Glaciation: Centuries Upon Centuries of Frigid Cold

Whatever the true story of recent human evolution, modern people like the Neanderthals emerged at the start of the last great glacial epoch, called the Würm. Several oceanic, atmospheric, and astronomical phenomena produced a major cooling of the climate 110,000 years ago. This period ended barely 10,000 years ago. The cooling was caused by changes in the earth's orbit around the sun and in the tilt of the globe's axis of rotation, magnified by profound modifications in polar marine currents and in variations in the atmosphere's carbon dioxide content.

In the Northern Hemisphere, landscapes and the contours of land and sea were completely transformed by these phenomena. During the coldest time, 18,000 years ago, when the ice reached a thickness of about two miles in places, the increase in the ice sheets' extent caused the sea level to drop by almost four hundred feet. As a result, a continental bridge more than six hundred miles long joined Alaska to Siberia.

During this period, in western Europe, the Neanderthals

Some 18,000 years ago an icecap covered Scandinavia, the Baltic Sea, and part of the North Sea and the British Isles, as well as the plains of North Germany.

experienced very varied living conditions. The climate was sometimes cold and dry, sometimes mild and humid. At times the extreme polar harshness caused the ground to be frozen all year round. The Neanderthals lived in a variety of environments, occupying open-air camps or rock shelters and caves. While some lived on the steppes or near the great glaciers, where reindeer, mammoths, and woolly rhinoceroses grazed, others inhabited the tundra of northern Europe.

The discovery of several complete Neanderthal skeletons has made it possible for a very accurate profile to be drawn. Neanderthals were quite small and thickset with short legs. They had large receding heads, attenuated at the rear. Their faces were marked by a prominent nose, receding cheekbones, and a low brow with a strong bony ridge over the eye sockets. Of course, other details—the shape of their lips, the color of their skin, the amount of hair on their faces and heads—is still not known.

Because of their powerful musculature, their strong build, and, particularly, the size of their heads, the Neanderthals have long personified the bestiality that

During the last great glaciation Europe was covered by a mosaic of steppes, tundras, grasslands, and forests. Their extent varied with the fluctuations in the climate. The Neanderthals often lived in the coldest areas, alongside mammoths, cave bears, and woolly rhinoceroses, or on the tundra with reindeer and musk-oxen. Other Neanderthals were to be found in the steppes and grasslands with bison, aurochs (wild cattle), and horses.

The mammoth captured the imagination of our ancestors, judging from their paintings and engravings. In its most recent form, *Mammuthus primigenius*, the woolly mammoth, was covered in a thick shaggy coat with hairs that could reach three feet in length. It appeared during the penultimate glaciation around 300,000 years ago and suddenly disappeared some 12,000 years ago. (However, some "dwarf" mammoths survived on Wrangel Island, north of Siberia, until about 3000 years ago.) Having spread throughout Europe and Asia, in a fairly small form close to that of the present-day Asian elephant, the mammoth colonized North America by crossing the land bridge that joined Alaska to Siberia during the coldest periods of the last Ice Age.

some liked to imagine existed in prehistoric humans. However, their preoccupation with the spiritual, their highly worked tools (resulting from a technique that could not have been learned without the help of language), and their hunting methods have provoked many people today to reassess their image of the Neanderthals. Besides, their brain was so large that it sometimes exceeded the average cranial capacity of present-day humans. They are associated with a very developed industry of worked stones, known as Mousterian. They shaped tools from flakes to make points and scrapers, retouching them into various forms.

Ritual Cannibalism?

Like the first modern humans, the Neanderthals have left behind in their burials evidence of their attitude towards death. They certainly were the first to bury their dead. Paleontologists are always delighted to discover such burials because—apart from exceptional cases like Lucy, from Hadar in Ethiopia, of which forty percent of the skeleton was found, and the recently discovered Preneanderthal of Altamura in Italy—it is only in this way that they are able to study complete skeletons. The remains are well preserved because they had been protected against scavenging animals.

Some bodies have been found in a bent position: For example, the rock shelter of La Ferrassie in the Dordogne contains a family burial of the presumed parents and six children, including a fetus and two newborn infants.

While traces of such burials are common in France, others are known in Italy, the Middle East, and further east in Central Asia. At the famous cave of Shanidar, Iraq, nine individuals were found at different depths. The oldest dated back 70,000 years.

A variety of tools and worked objects are sometimes buried with them. At Shanidar flowers—including cornflowers, thistles, and hollyhocks—known for

Burials made by Upper Paleolithic people express their attitude towards death. Only a few individuals in the group were worthy of this special treatment, and they were buried in a variety of ways. At the Magdalenian burial of St.-Germain-la-Rivière in the Gironde, France, the deceased's head was protected by stone slabs (below). The bodies buried in the Grimaldi caves in northwestern Italy (pages 116 and 117 top) were arranged very differently.

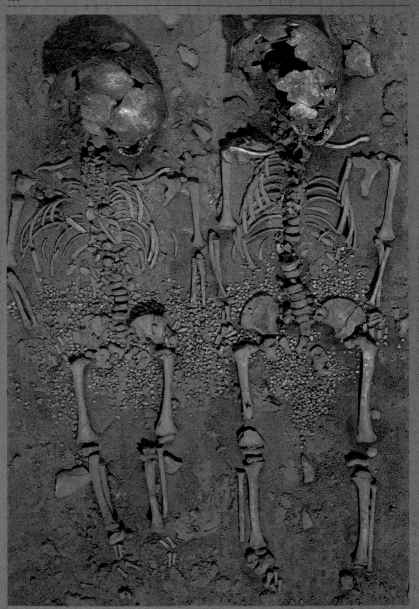

Opposite and left: Burials at the Grimaldi caves. Below left: A Neanderthal man was discovered at La Ferrassie in the Dordogne in 1909. He lay buried at the bottom of a pit dug in the ground.

This drawing (above), which shows the position of the La Ferrassie body as it was found, was made by Abbé Breuil in 1909.

their medicinal properties, were placed in the grave. They may have been deposited there to honor the deceased for his or her healing powers or simply to provide decoration.

All these burials show evidence of some sort of ritual taking place, as well as a concern with an afterlife.

Some of these burials may have been linked to cannibalistic practices. In the cave of Hortus in the Hérault in southern France, several human bones have been discovered mixed with food debris. At Krapina in Croatia, many of the human bones were smashed and split.

Although the evidence has been disputed, it is possible that these two examples reveal the existence of ritual cannibalism, which is still found today among some peoples in Africa and Melanesia.

While the Magdalenian culture in Europe (about 17,000 to 10,000 BC) has left numerous traces of dwellings in the open air or in natural shelters, during the same period the hunters of central Europe and as far as Siberia erected huge settlements of mammoth bones and tusks. Mezhirich in Ukraine, which was occupied about 15,000 years ago, is a particularly spectacular example (above). The lowest layer of each hut contained almost a hundred mammoth lower jaws stacked one on top of another (reconstruction, left).

Homo Sapiens Sapiens: The Invader from the East Took Over

For reasons that still remain a mystery, the Neanderthals, whose last representatives are associated with the Châtelperronian industry, were to disappear suddenly around 35,000 years ago. They then gave way to anatomically modern humans, who no longer had any of the typical physical characteristics of the Neanderthals. Most paleoanthropologists now accept that modern humans did not evolve from the Neanderthals in Europe. For their part, prehistorians have observed a clear break in the evolution of knapping techniques, between the Châtelperronian stone tool kits made by the last Neanderthals and the more recent tool kits from the Aurignacian culture.

Further evidence of a break is provided by the first manifestations of figurative art and by a more elaborate boneworking

Numerous decorated, engraved, or sculpted objects, very often depicting animals, as in the case of these spearthrowers, express the rich artistic sense of Magdalenian hunters. Weapons of this type, with a hook at one end, enabled finely worked spearpoints made from reindeer antler to be thrown with a rapid movement of the arm and wrist.

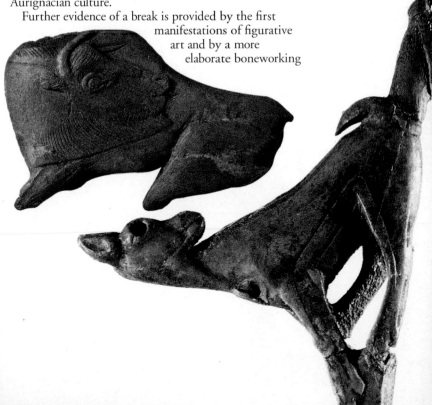

technology seen in the emergence of spear points and those carved pieces of reindeer antler known as perforated batons.

These biological and cultural developments suggest that the Neanderthals were replaced by modern humans who invaded western Europe from the East. If this was the case, there can be no doubt that the two populations crossbred. But who were these invaders? To judge by their physical features—an elevated skull with a high brow, a face with prominent cheekbones and a protruding chin— they were the first modern humans, *Homo sapiens sapiens*, direct ancestors of the Cro-Magnons.

The Cro-Magnon people, hunter-gatherers who appeared in Europe 35,000 years ago, bear such a close resemblance to modern humans that, it has been said, if they walked around London in a suit, it is unlikely that anybody would comment on their appearance.

Cro-Magnon: Cave Art

One of the most striking aspects of the Cro-Magnon age is the development of aesthetic feelings, which are closely linked to religion or magic.

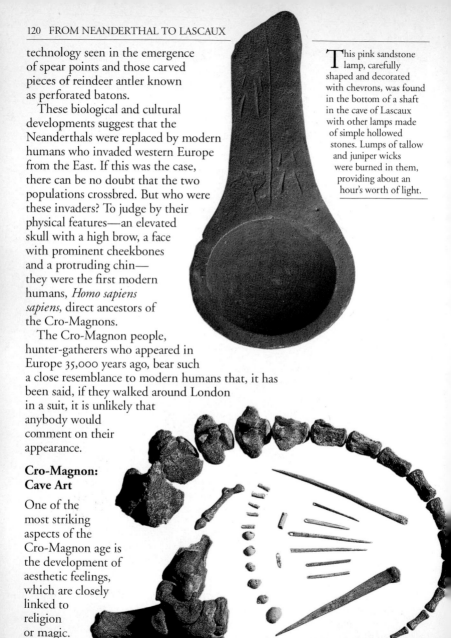

This pink sandstone lamp, carefully shaped and decorated with chevrons, was found in the bottom of a shaft in the cave of Lascaux with other lamps made of simple hollowed stones. Lumps of tallow and juniper wicks were burned in them, providing about an hour's worth of light.

On 12 September 1940 four youngsters from the village of Montignac in the Dordogne discovered the entrance to the cave of Lascaux, which was to become one of the centerpieces of prehistoric art. This cave, a monumental underground museum about 17,000 years old, contained more than six hundred rock paintings and almost fifteen hundred engravings. Abbé Henri Breuil was called to the site to authenticate the paintings. He made the first tracings (left) and devoted his life to depicting and studying Upper Paleolithic cave art.

The people of the last Ice Age developed boneworking to a high level. Examples of their craft include awls, which enabled them to sew skins, and necklaces made by stringing various objects onto tendons (opposite).

Artistic expression is found in engravings, paintings, sculptures, and ornaments. Ocher, whose color varied from red to yellow, was widely used. The Cro-Magnon people seem to have attached a religious meaning to red ocher, on which they sometimes laid their dead. It is also possible that they tinted animal hides and painted their bodies. Certainly, bone, ivory, and shells were worked into decorative objects, pendants, and necklaces.

Their first drawings, mostly engravings, rarely painted, were still crude and generally clumsy. Figurative art in the shape of more elaborate engravings and paintings gradually emerged from the first scrawls. Throughout Europe this interest in art found expression in small statuettes of young girls with elegant features and of

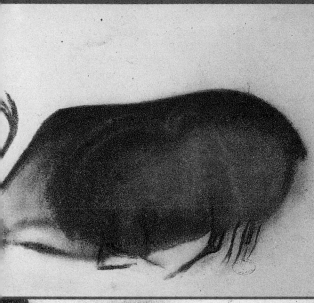

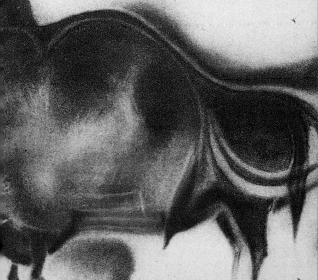

Underground Art

The cave art of the numerous underground sanctuaries of France (especially in Aquitaine and the Pyrenees) and Spain (such as Altamira, where one of the very first decorated caves was discovered in 1879) share a common theme. Countless depictions of animals—horses, aurochs, bison, reindeer, and mammoths hunted at the time—are found during the last 15,000 years (from about 25,000 to 10,000 years ago) of the Upper Paleolithic. This era produced art of extraordinary skill and achievement. There is evidence that the people of this period took part in cult ceremonies in the depths of these dark caves, in which they stayed for only a short time.

Left: Tracings of Upper Paleolithic reindeer (above) and mammoth and bison (below) from Font de Gaume, Dordogne, by Abbé Breuil. His tracing of the famous painted ceiling at Altamira, Spain, is on pages 124–5.

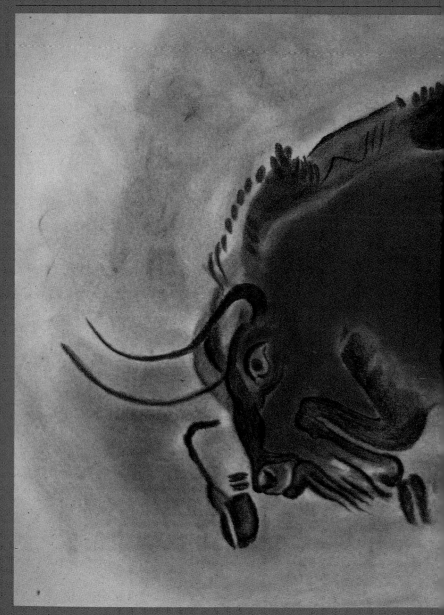

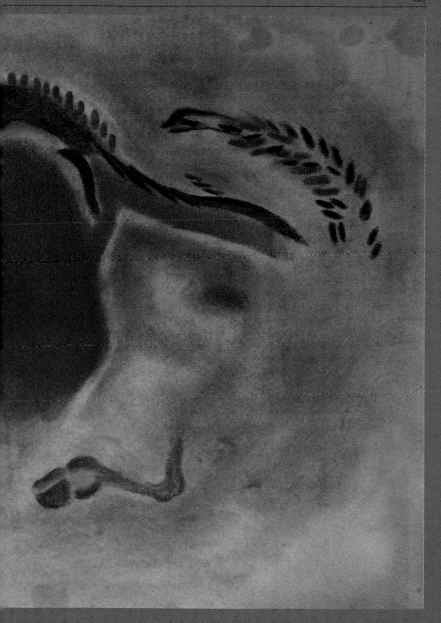

fleshy matrons, often with extremely stylized features; these are known as Venuses.

The people of this culture known as the reindeer age (the last Ice Age), who lived exclusively from hunting, fishing, and gathering, brought cave art to its zenith between 17,000 and 9000 years ago; one of the best examples of their art is the cave of Lascaux, France. The artists tended to work in relatively accessible caves, though they sometimes ventured far underground, as at the cave of Niaux in the Ariège, which could only be reached after a long and sometimes perilous journey.

Apart from a few rare exceptions, these underground sanctuaries, where magical rites of sorcery and initiation linked to the world of hunting were performed, have only been found in France and Spain, on both sides of the Pyrenees. Cave art was to disappear abruptly nearly 10,000 years ago, just as the last great Ice Age ended. It then took just a few millennia for humans to domesticate plants and animals, invent pottery, and discover metallurgy.

With the emergence of consciousness —which sets us apart from all other forms of life—humans started to question where we fit into the celestial mechanism. Whatever our destiny might be, Albert Einstein (below) was continually amazed at our ability to understand the world around us.

Page 128: Neanderthals, as depicted by Zdenek Burian.

And What Does Tomorrow Hold? Dawn or Twilight?

Theoretically, no plant or animal in the future will remain as it is now. Evolution is a constant process that continues to take place today. But humans are peculiar animals. For a few thousand years human biological evolution has slowed down considerably, while the pace of cultural evolution continues to speed up. A few hundred years from now it is likely that humanity will almost totally free itself from the biological laws that have governed its evolution.

The processes underlying biological evolution —those that led us from being apes to being humans—work extremely slowly in comparison with those underlying cultural evolution. Indeed, the mechanisms of natural selection (relying on the success of individuals) do not enable acquired characteristics to be passed from one generation to another. Biological transmission is essentially carried out by small alterations that are barely perceptible over a short period of time.

Cultural evolution, however, which relies on the success of the group, causes changes to take place very rapidly, because the transmission process is cumulative in the sense that each generation passes on—either through speech or through the written word—its acquired knowledge to the next.

Thanks to our exceptional intelligence, which is linked to the ongoing increase in the size of our brains over tens of thousands of years, humans possess the capability to mold the future with complete awareness. In this respect humans are unique in the animal kingdom. However, at the same time, we are the only animals who have become a threat to ourselves.

It is impossible to foresee which force will prevail in the future: the force of cultural evolution that would lead us towards a new form of *Homo sapiens* or the force of destruction that would eliminate us in order to restore the balance of nature before nature itself is engulfed in a few billion years by a gigantic ball of fire.

DOCUMENTS

The Descent of Man

In 1859 Darwin published his groundbreaking Origin of Species, *in which he set out the theory of evolution based on natural selection. Twelve years later, in 1871, he applied the theory to human beings in* The Descent of Man.

T his caricature of Darwin appeared the year he published *The Descent of Man*. People were still shocked by his ideas.

Natural Selection

We have now seen that man is variable in body and mind; and that the variations are induced, either directly or indirectly, by the same general causes, and obey the same general laws, as with the lower animals. Man has spread widely over the face of the earth, and must have been exposed, during his incessant migrations, to the most diversified conditions.... The early progenitors of man must also have tended, like all other animals, to have increased beyond their means of subsistence; they must therefore occasionally have been exposed to a struggle for existence, and consequently to the rigid law of natural selection. Beneficial variations of all kinds will thus, either occasionally or habitually, have been preserved, and injurious ones eliminated. I do not refer to strongly-marked deviations of structure, which occur only at long intervals of time, but to mere individual differences....

Man in the rudest state in which he now exists is the most dominant animal that has ever appeared on the earth. He has spread more widely than any other highly organised form; and all others have yielded before him. He manifestly owes this immense superiority to his intellectual faculties, his social habits, which lead him to aid and defend his fellows, and to his corporeal structure. The supreme importance of these characters has been proved by the final arbitrament of the battle for life....

On the Birthplace and Antiquity of Man

We are naturally led to enquire where was the birthplace of man at that stage of descent when our progenitors diverged from the Catarhine [monkey] stock. The fact that they belonged to this stock clearly shews that they inhabited the Old World; but not Australia nor any oceanic island, as we may infer from the laws of geographical distribution. In each great region of the world the living mammals are closely related to the extinct species of the same region. It is therefore probable that Africa was formerly inhabited by extinct apes closely allied to the gorilla and chimpanzee; and as these two species are now man's nearest allies, it is somewhat more probable that our earliest progenitors lived on the African continent than elsewhere....

The Principle of Evolution

The main conclusion arrived at in this work, and now held by many naturalists who are well competent to form a sound judgment, is that man is descended from some less highly organised form. The grounds upon which this conclusion rests will never be shaken, for the close similarity between man and the lower animals in embryonic development, as well as in innumerable points of structure and constitution, both of high and of the most trifling importance, —the rudiments which he retains, and the abnormal reversions to which he is occasionally liable,—are facts which cannot be disputed. They have long been known, but until recently they told us nothing with respect to the origin of man.

Now when viewed by the light of our knowledge of the whole organic world, their meaning is unmistakable. The great principle of evolution stands up clear and firm, when these groups of facts are considered in connection with others, such as the mutual affinities of the members of the same group, their geographical distribution in past and present times, and their geological succession. It is incredible that all these facts should speak falsely. He who is not content to look, like a savage, at the phenomena of nature as disconnected, cannot any longer believe that man is the work of a separate act of creation. He will be forced to admit that the close resemblance of the embryo of man to that, for instance, of a dog—the construction of his skull, limbs, and whole frame, independently of the uses to which the parts may be put, on the same plan with that of other mammals —the occasional reappearance of various structures, for instance of several distinct muscles, which man does not normally possess, but which are common to the Quadrumana—and a crowd of analogous facts—all point in the plainest manner to the conclusion that man is the co-descendant with other mammals of a common progenitor.

Variation

We have seen that man incessantly presents individual differences in all parts of his body and in his mental faculties. These differences or variations seem to be induced by the same general causes, and to obey the same laws as with the lower animals. In both cases similar laws of inheritance prevail. Man tends to increase at a greater rate than his means of subsistence; consequently he is occasionally subjected to a severe struggle for existence, and natural selection will have effected whatever lies within its scope. A succession of strongly-marked variations of a similar nature are by no means requisite; slight fluctuating differences in the individual suffice for the work of natural selection. We may feel assured that the inherited effects of the long-continued use or disuse of parts will have done much in the same direction with natural selection. Modifications formerly of importance, though no longer of any special use, will be long inherited. When one part is modified, other parts will change through the principle of correlation, of which we have instances in many curious cases of correlated monstrosities. Something may be attributed to the direct and definite action of the surrounding conditions of life, such as abundant food, heat, or moisture; and lastly, many characters of slight physiological importance, some indeed of consider-able importance, have been gained through sexual selection....

The Origin of Man

By considering the embryological structure of man,—the homologies which he presents with the lower animals,—the rudiments which he retains,—and the reversions to which he is liable, we can partly recall in imagination the former condition of our early progenitors; and can approximately place them in their proper position in the zoological series. We thus learn that man is descended from a hairy quadruped, furnished with a tail and pointed ears, probably arboreal in its habits, and an inhabitant of the Old World....

Intellectual Ability

The greatest difficulty which presents itself, when we are driven to the above conclusion on the origin of man, is the high standard of intellectual power and of moral disposition which he has attained. But every one who admits the general principle of evolution, must see that the mental powers of the higher animals, which are the same in kind with those of mankind, though so different in degree, are capable of advancement. Thus the interval between the mental powers of one of the higher apes and of a fish, or between those of an ant and scale-insect, is immense. The development of these powers in animals does not offer any special difficulty; for with our domesticated animals, the mental faculties are certainly variable, and the variations are inherited. No one doubts that these faculties are of the utmost importance to animals in a state of nature. Therefore the conditions are favourable for their development through natural selection. The same conclusion may be extended to man; the intellect must have been all-important to him, even at a very remote period, enabling him to use language, to invent and make weapons, tools, traps, etc.; by which means, in combination with his social habits, he long ago became the most dominant of all living creatures.

A great stride in the development of the intellect will have followed, as soon as, through a previous considerable advance, the half-art and half-instinct of language came into use; for the continued use of language will have reacted on the brain, and produced an inherited effect; and this again will have reacted on the improvement of language....

Morality and Religion

The belief in God has often been advanced as not only the greatest, but the most complete of all the distinctions between man and the lower animals. It is however impossible, as we have seen, to maintain that this belief is innate or instinctive in man. On the other hand a belief in all-pervading spiritual agencies seems to be universal;

and apparently follows from a considerable advance in the reasoning powers of man, and from a still greater advance in his faculties of imagination, curiosity and wonder.

I am aware that the assumed instinctive belief in God has been used by many persons as an argument for His existence. But this is a rash argument, as we should thus be compelled to believe in the existence of many cruel and malignant spirits, possessing only a little more power than man; for the belief in them is far more general than of a beneficent Deity. The idea of a universal and beneficent Creator of the universe does not seem to arise in the mind of man, until he has been elevated by long-continued culture.

He who believes in the advancement of man from some lowly-organised form, will naturally ask how does this bear on the belief in the immortality of the soul.... Few persons feel any anxiety from the impossibility of determining at what precise period in the development of the individual, from the first trace of the minute germinal vesicle to the child either before or after birth, man becomes an immortal being; and there is no greater cause for anxiety because the period in the gradually ascending organic scale cannot possibly be determined.

I am aware that the conclusions arrived at in this work will be denounced by some as highly irreligious; but he who thus denounces them is bound to shew why it is more irreligious to explain the origin of man as a distinct species by descent from some lower form, through the laws of variation and natural selection, than to explain the birth of the individual through the laws of ordinary reproduction. The birth both of the species and of the individual are equally parts of the grand sequence of events, which our minds refuse to accept as the result of blind chance. The understanding revolts at such a conclusion, whether or not we are able to believe that every slight variation of structure,—the union of each pair in marriage,—the dissemination of each seed,—and other such events, have all been ordained for some special purpose.

Conclusion

The main conclusion arrived at in this work, namely that man is descended from some lowly-organised form will, I regret to think, be highly distasteful to many persons. But there can hardly be a doubt that we are descended from barbarians....

Man may be excused for feeling some pride at having risen, though not through his own exertions, to the very summit of the organic scale; and the fact of his having thus risen, instead of having been aboriginally placed there, may give him hopes for a still higher destiny in the distant future. But we are not here concerned with hopes or fears, only with the truth as far as our reason allows us to discover it. I have given the evidence to the best of my ability; and we must acknowledge, as it seems to me, that man with all his noble qualities, with sympathy which feels for the most debased, with benevolence which extends not only to other men but to the humblest living creature, with his god-like intellect which has penetrated into the movements and constitution of the solar system—with all these exalted powers—Man still bears in his bodily frame the indelible stamp of his lowly origin.

Charles Darwin
The Descent of Man
1871

The Scopes Trial

In 1925 Americans watched a spectacular confrontation between ancient and modern, between fundamentalists and evolutionists. The trial brought against John Thomas Scopes, a Tennessee schoolteacher, focused attention on whether humans are a product of a literal biblical creation or whether we evolved from apes.

George W. Rappelyea [a mining engineer]…saw in a newspaper that Chattanooga had given up its plans to start a case to test the Butler Act. He got an idea, and he telephoned F. E. Robinson, local druggist and head of the county board of education, and Walter White, county superintendent of schools. He argued earnestly with them. The next day he was at them again. They gave in. Then Rappelyea sent for John Thomas Scopes and asked him to come down to Robinson's drugstore.…

John Scopes was a guileless young man, with blue, contemplative eyes. Only 24 years old, he had graduated from the University of Kentucky the preceding year and had come to the high school at Dayton as science teacher and football coach. His local popularity was very great. Here was the man Rappelyea wanted. Scopes was drawn into the discussion, and found himself observing that nobody could teach biology without using the theory of evolution. Being the person he was, he was trapped. Rappelyea said, "You have been violating the law."

"So has every other teacher," said Scopes. "This is the official textbook.".…

On May 7 John Scopes was arrested. … It was charged that on April 24 he had taught the theory of evolution to his class.…

If Governor Peay and others saw the Scopes trial partly as political speech and partly as pious gesture, such men as George Rappelyea, Sue Hicks [a male lawyer] and Dayton merchants viewed it as a civic promotion. This could really put Dayton in the headlines, on the map. It could bring a lot of business to local stores. But it needed celebrities. Who better than William Jennings Bryan, the one man of world reputation in the fundamentalist movement?…

Henry Fairfield Osborn [director of the American Museum of Natural History] said that the trial would do great good by clarifying the issues.... *The Real Question Is, Did God Use Evolution in His Plan?* Beside this question all of the others—"such as personal rights, rights of opinion, rights of free speech, constitutional rights, education liberty" were insignificant....

Clarence Darrow [the chief defense attorney] agreed. Although he thought the references to God were impertinent, he did regard the truth of evolution as the main issue. His whole approach to the trial was to establish this truth by means of scientific testimony....

It was not enough, said Malone [another defense attorney] in presenting the defense theory, for the state to show that Scopes had taught the theory of evolution; the state must show in addition that Scopes had "also, and at the same time, denied the theory of creation as set forth in the Bible." In brief, what did the Butler Act forbid: Teaching that man had descended from a lower order of animals? Teaching a story of human origins that conflicted with Genesis? Or both? The state held to the first construction of the law; the defense, to the third. They wrangled about it throughout the trial—and beyond....

Darrow began quietly, asking if Bryan [the prosecution counsel, here in the witness stand] had not "given considerable study to the Bible." Bryan admitted it. Then Darrow began his ruthless efforts to make Bryan admit that the Bible could not always be taken literally, that it was sometimes vague, that the Butler Act was fatally indefinite when it forbade the teachings of "any theory that denies the story of the Divine Creation of man as taught in the Bible."...

Darrow kept worrying at the witness about the dates that Bishop Ussher, calculating from the ages of the various prophets, had assigned to Scriptural episodes. The Bishop had computed the date of Creation as 4004 BC. He had been even more specific: this happy event had occurred on October 23 at 9 AM....

Bryan wriggled and writhed, but Darrow kept pressing him. And eventually Bryan gave answers.

Did Bryan believe that all of the species on the earth had come into being in the 4200 years, by the Bishop's dating, since the Flood occurred? Yes, said Bryan finally, he did believe it.

Didn't Bryan know that many civilizations had existed for more than 5000 years? Said Bryan: "I have never felt a great deal of interest in the effort that has been made to dispute the Bible by the speculations of men, or the investigations of men...."

By this time Bryan's self-esteem was suppurating, and his wits had entirely deserted him. Having discredited himself with everybody who did not believe in the literal truth of the Bible, he now destroyed himself with those who did. It took one deft question by Darrow, and a six-word reply.

Darrow asked: "Do you think the earth was made in six days?"

Bryan: "Not six days of twenty-four hours...."

Bryan said that the defense lawyers had "no other purpose than ridiculing every Christian who believes in the Bible."

Darrow said directly to Bryan: "We have the purpose of preventing bigots and ignoramuses from controlling the education of the United States and you know it—and that is all."

Ray Ginger
Six Days or Forever?, 1958

Australopithecus Africanus: The Southern Ape from South Africa

In 1925 Raymond Dart (opposite) published the results of his investigation into a fossil skull found at Taung (or Taungs) in South Africa. With great audacity he proposed a completely new view of human evolution based on this child's skull. He named the species to which this child belonged Australopithecus africanus *and saw it as the missing link between apes and humans. His theories met with a mixed reception. Until the discovery of Lucy in 1974, this species was thought to be the oldest-known human ancestor.*

Towards the close of 1924, Miss Josephine Salmons, student demonstrator of anatomy in the University of the Witwatersrand, brought to me the fossilised skull of a cercopithecid monkey which, through her instrumentality, was very generously loaned to the Department for description by its owner, Mr. E. G. Izod, of the Rand Mines Limited. I learned that this valuable fossil had been blasted out of the limestone cliff formation—at a vertical depth of 50 feet and a horizontal depth of 200 feet—at Taungs, which lies 80 miles north of Kimberley on the main line to Rhodesia, in Bechuanaland, by operatives of the Northern Lime Company. Important stratigraphical evidence has been forthcoming recently from this district concerning the succession of stone ages in South Africa ... and the feeling was entertained that this lime deposit, like that of Broken Hill in Rhodesia, might contain fossil remains of primitive man.

I immediately consulted Dr. R. B. Young, professor of geology in the University of the Witwatersrand, about the discovery, and he, by a fortunate coincidence, was called down to Taungs almost synchronously to investigate geologically the lime deposits of an adjacent farm. During his visit to Taungs, Prof. Young was enabled, through the courtesy of Mr. A. F. Campbell, general manager of the Northern Lime Company, to inspect the site of the discovery and to select further samples of fossil material for me from the same formation. These included a natural cercopithecid endocranial cast, a second and larger cast, and some rock fragments disclosing portions of bone. Finally, Dr. Gordon D. Laing, senior lecturer in anatomy, obtained news, through his friend Mr.

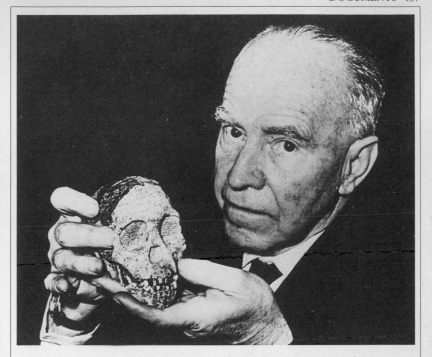

Ridley Hendry, of another primate skull from the same cliff. This cercopithecid skull, the possession of Mr. De Wet, of the Langlaagte Deep Mine, has also been liberally entrusted by him to the Department for scientific investigation....

In manipulating the pieces of rock brought back by Prof. Young, I found that the larger natural endocranial cast articulated exactly by its fractured frontal extremity with another piece of rock in which the broken lower and posterior margin of the left side of a mandible was visible. After cleaning the rock mass, the outline of the hinder and lower part of the facial skeleton came into view. Careful development of the solid limestone in which it was

embedded finally revealed the almost entire face....

It was apparent when the larger endocranial cast was first observed that it was specially important, for its size and sulcal pattern revealed sufficient similarity with those of the chimpanzee and gorilla to demonstrate that one was handling in this instance an anthropoid and not a cercopithecid ape. Fossil anthropoids have not hitherto been recorded south of the Fayüm in Egypt, and living anthropoids have not been discovered in recent times south of Lake Kivu region in Belgian Congo, nearly 2000 miles to the north, as the crow flies.

All fossil anthropoids found hitherto have been known only from mandibular or maxillary fragments, so far as crania are concerned, and so the general appearance of the types they represented has been unknown; consequently, a condition of affairs where virtually the whole face and lower jaw, replete with teeth, together with the major portion of the brain pattern, have been preserved, constitutes a specimen of unusual value in fossil anthropoid discovery. Here, as in *Homo rhodesiensis*, Southern Africa has provided documents of higher primate evolution that are amongst the most complete extant.

Apart from this evidential completeness, the specimen is of importance because it exhibits an extinct race of apes *intermediate between living anthropoids and man*.... It is obvious, meanwhile, that it represents a fossil group distinctly advanced beyond living anthropoids in those two dominantly human characters of facial and dental recession on one hand, and improved quality of the brain on the other. Unlike Pithecanthropus, it does not represent an ape-like man, a caricature of precocious hominid failure, but a creature well advanced beyond modern anthropoids in just those characters, facial and cerebral, which are to be anticipated in an extinct link between man and his simian ancestor. At the same time, it is equally evident that a creature with anthropoid brain capacity, and lacking the distinctive, localised temporal expansions which appear to be concomitant with and necessary to articulate man, is no true man. It is therefore logically regarded as a man-like ape.

I propose tentatively, then, that a new family of *Homo-simiadae* be created for the reception of the group of individuals which it represents, and that the first known species of the group be designated *Australopithecus africanus*,

in commemoration, first, of the extreme southern and unexpected horizon of its discovery, and, secondly, of the continent in which so many new and important discoveries connected with the early history of man have recently been made, thus vindicating the Darwinian claim that Africa would prove to be the cradle of mankind.

Professor Raymond A. Dart
Nature
7 February 1925

The New Missing Link

Professor Raymond Dart, whose discovery of a "missing link" in South Africa has fallen like a bombshell on anthropological Europe, is well known to British anatomists. He is one of the many young medical graduates of Sydney University whose minds were bent towards research in anatomy by Professor J. T. Wilson, before this distinguished anatomist left Australia for Cambridge. In post-graduate days Professor Dart worked at University College, London, with Professor Elliot Smith, at the Royal College of Surgeons, and in research laboratories of the United States. He went to Johannesburg some three years ago to occupy the chair of anatomy in the University of Witwatersrand. In South Africa he found an earnest band of investigators opening up unexpected chapters in the early history of man. Old river-beds were yielding a sequence of stone implements almost as ancient as those of Europe. The remains of a fossil man had been found at Boskop, and soon after Professor Dart's arrival other human remains were found deep in the floor of a rock shelter on the southern coast of Cape Colony. These were examined by Professor Dart, and found to be of the Boskop race. Then came the discovery of Rhodesian man in the bowels of a limestone kopje in Rhodesia—a much older and more primitive type than that of Boskop. And now comes an equally remarkable discovery which is largely due to the initiative and perspicuity of Professor Dart.

In November last a lady demonstrator brought him the fossil skull of a monkey —a baboon. It came from a limestone quarry situated at Taungs, eighty miles to the north of Kimberley. There the quarrymen had worked their way 200 feet into a limestone bluff which rises 50 feet above the dry veldt of the surrounding country. The monkey's skull was blasted from the rock. Finding that his colleague, Dr. R. B. Young, professor of geology, was going to Taungs, he asked him to inquire into the source of the fossil bones in the quarry. Dr. Young arrived at the quarry in time to receive a mass of material just blasted from the base of the working face of the quarry. He gathered the fragments, and handed them to Professor Dart on his return.

In the mass of limestone Professor Dart found a cast which had formed within the brain cavity of a skull, and from the adjoining block he chiselled out the forehead and complete facial parts which went with the brain cast. The blasting had shivered and destroyed most of the cranial bones. From these fragments the discoverer reconstructed the being to which he has given the name *Australopithecus africanus.* So exact and clear are his drawings and his descriptions that those who have studied his preliminary account in *Nature*… have all the data placed at their disposal for coming to an independent opinion. Indeed, those who have charge of much larger collections of anthropoid and human skulls and brains than were at the disposal of Professor Dart have a somewhat unfair advantage over him. But with all these disadvantages against him his main conclusions are certain to stand.

He speaks of this new being as a "man-ape," and as standing "between living anthropoids and man." These are his expressions, but when we examine his text we find him quite alive to the fact that the animal he has brought before the scientific world is a man-like ape or anthropoid. The size and convolutionary pattern of its brain leaves one in no doubt of this matter. Many fossil fragments of higher anthropoid apes have been found on previous occasions in Europe and India, but this is the first time we have seen the complete face of one of them. In this case the animal is young: the first permanent molar teeth have cut and are coming into place; this happens in the gorilla and chimpanzee towards the end of the fourth year—two years earlier than in human children. The face in all its lineaments is that of an anthropoid; there are blended in it some features of the chimpanzee, others of the gorilla, and some which belong to neither. But of humanity there is no trace save in one respect—its jaws are smaller and its supraorbital ridges less developed than in a chimpanzee of a corresponding age. There is a reduction in jaw development, and such a reduction has certainly taken place in the evolution of man. Further, the milk canines are less pointed than are those of the young chimpanzee, and the interdental space in front of the upper canine is less.

Sir Arthur Keith
The British Medical Journal
14 February 1925

Finding ER-1470, A *Homo Habilis* Skull

In 1972 Richard E. Leakey (below) brought world attention to a skull discovered at Koobi Fora near Lake Turkana in Kenya. Known by its catalogue number at the National Museum of Kenya, ER-1470 was important because it was the oldest, most complete representative of Homo habilis *to have been found by that time.*

The year 1972 was an eventful one. Meave gave birth to our first daughter, Louise, on 21 March; the best known of all our discoveries from Koobi Fora, the skull "1470," was made in July; and in October my father died....

The skull "1470," the earliest evidence we had for *Homo* at Koobi Fora, was discovered by Bernard Ngeneo who, although he had only joined Kamoya's search team the previous year, quickly became an accomplished fossil hunter....

The 1972 discovery of "1470" has had tremendous publicity and is certainly the best-known fossil from Koobi Fora. When found, however, it caused no real excitement other than the usual good feeling that another hominid had been discovered. I was away in Nairobi at the time, but when I visited the site several days later on 27 July, everything was just as Bernard had found it, nothing had been disturbed. The specimen was badly broken and many fragments of light-coloured fossil bone were lying on the surface of a steep-sided ravine. None of the fragments was more than an inch long, but some were readily recognizable as being part of a hominid cranium. One good thing that was immediately apparent was that some were obviously from the back of the skull, others from the top, some from the sides, and there were even pieces of the very fragile facial bones. This indicated that there was a chance that we might eventually find enough pieces to reconstruct a fairly complete skull. It was clear, however, that a major sieving operation was required to recover other fragments that might be lying buried in the top few inches of soil or which had been washed down the steep slope. This sieving operation was not begun until a fortnight later and it continued over many weeks.

A number of fragments were collected in the first few days of sieving. On the fifth day, Meave, Bernard Wood (a friend who had been with me on several previous expeditions) and I flew to the site to help. At lunch time we returned to Koobi Fora with a number of fragments and after eating and a welcome swim we retired to the shady verandah of our house to examine the pieces. Meave carefully washed the fragments and laid them out on a wooden tray to dry in the sun and before long we were ready to begin to find which pieces could be joined to others. In no time at all, several of the bigger pieces fitted together and we realized that the fossil skull had been large, certainly larger than the small-brained *Australopithecus* such as we had found in 1969 and 1970. By the end of that exciting afternoon, we knew that we could go no further with the reconstruction without more pieces from the sieving.

Over the next few weeks more and more pieces were found in the sieving and Meave slowly put the fragments together. Gradually a skull began to take shape and we began to get a rough idea of its size. It was larger than any of the early fossil hominids that I had seen but the question was, how large was the brain? We decided to attempt a crude guess. Beginning by carefully filling the gaps in the vault with Plasticine and sticky tape, we then filled the vault with beach sand and measured the volume of sand in a rain gauge. By a most complicated conversion we came up with a volume of just under 800 cubic centimetres. The actual value for the brain size of "1470" has since been established by accurate methods as 775 cubic centimetres, so we were very close. This was fantastic new information. We now had an early fossil human skull with a brain size considerably larger than anything that had been found before of similar antiquity. Also, we had found some limb bones. At the time we believed that the skull must be older than 2.6 million years—this being based upon the dating of the KBS tuff and the assurances that we had from John Miller [a British geophysicist at Cambridge] and Frank Fitch [a colleague] to the effect that this was a good date. It turned out that we were wrong by at least half a million years but this we only learned much later....

The whole question of whether a skull should be called *Homo* or something else is a matter of definition. None of the fossils that we find are labelled. We give them names for our own convenience. We have to judge whether "X" looks more like "Y" than "Z" and this decision is often made more difficult because "X," "Y" and "Z" are incomplete. I called this particular skull *Homo* because I believed it to be more like other fossils that had been called *Homo* than it was to those called *Australopithecus*. More importantly, "1470" has a brain size which is considerably bigger than any of the known fossils of *Australopithecus* and this is, in my opinion, very significant.

The intelligence we have, along with our technology and culture, all stems from some event way back in time when it was advantageous to be larger brained. My interest in early *Homo* is nothing more than a desire to determine exactly when the brain began to increase in size and there is no doubt, even after the revision of the dating, that "1470" is one of the earliest examples of a large-brained hominid.

One Life: Richard E. Leakey:
An Autobiography
1983

The Discovery of Lucy

On 30 November 1974 in a gully at Hadar, Ethiopia, Donald Johanson (below) made a unique find in the history of hominid fossil collecting. Here he recounts his first encounter with Lucy, as she soon became known, and tells of the excitement he felt at finding a hominid dating back approximately 3.1 million years in such a good state of preservation.

As a paleoanthropologist—one who studies the fossils of human ancestors—I am superstitious. Many of us are, because the work we do depends a great deal on luck. The fossils we study are extremely rare, and quite a few distinguished paleoanthropologists have gone a lifetime without finding a single one. I am one of the more fortunate. This was only my third year in the field at Hadar, and I had already found several. I know I am lucky, and I don't try to hide it. That is why I wrote "feel good" in my diary. When I got up that morning I felt it was one of those days when you should press your luck. One of those days when something terrific might happen....

At Hadar, which is a wasteland of bare rock, gravel and sand, the fossils that one finds are almost all exposed on the surface of the ground. Hadar is in the center of the Afar desert, an ancient lake bed now dry and filled with sediments that record the history of past geological events. You can trace volcanic-ash falls there, deposits of mud and silt washed down from distant mountains, episodes of volcanic dust, more mud, and so on. Those events reveal themselves like layers in a slice of cake in the gullies of new young rivers that recently have cut through the lake bed here and there. It seldom rains at Hadar, but when it does it comes in an overpowering gush—six months' worth overnight. The soil, which is bare of vegetation, cannot hold all that water. It roars down the gullies, cutting back their sides and bringing more fossils into view.

[Tom] Gray and I parked the Land-Rover on the slope of one of those gullies. We were careful to face it in such a way that the canvas water bag that was hanging from the side mirror was in the shade. Gray plotted the locality on the map. Then we got out and began doing what most members of the expedition spent a great deal of their time doing: we began surveying, walking slowly about, looking for exposed fossils....

Tom and I surveyed for a couple of hours. It was now close to noon, and the temperature was approaching 110. We hadn't found much: a few teeth of the small extinct horse *Hipparion*; part of the skull of an extinct pig; some antelope molars; a bit of a monkey jaw. We had large collections of all these things already, but Tom insisted on taking these also as added pieces in the overall jigsaw puzzle of what went where.

"I've had it," said Tom. "When do we head back to camp?"

"Right now. But let's go back this way and survey the bottom of that little gully over there."

The gully in question was just over the crest of the rise where we had been working all morning. It had been thoroughly checked out at least twice before by other workers, who had found nothing interesting. Nevertheless, conscious of the "lucky" feeling that had been with me since I woke, I decided to make that small final detour. There was virtually no bone in the gully. But as we turned to leave, I noticed something on the ground partway up the slope.

"That's a bit of a hominid arm," I said.

"Can't be. It's too small. Has to be a monkey of some kind."

We knelt to examine it.

"Much too small," said Gray again. I shook my head. "Hominid."

"What makes you so sure?" he said.

"That piece right next to your hand. That's hominid too."

"Jesus Christ," said Gray. He picked it up. It was the back of a small skull. A few feet away was part of a femur: a

thighbone. "Jesus Christ," he said again. We stood up, and began to see other bits of bone on the slope: a couple of vertebrae, part of a pelvis—all of them hominid. An unbelievable, impermissible thought flickered through my mind. Suppose all these fitted together? Could they be parts of a single, extremely primitive skeleton? No such skeleton had ever been found—anywhere.

"Look at that," said Gray. "Ribs."

A single individual?...

That afternoon everyone in camp was at the gully, sectioning off the site and preparing for a massive collecting job that ultimately took three weeks. When it was done, we had recovered several hundred pieces of bone (many of them fragments) representing about forty percent of the skeleton of a single individual. Tom's and my original hunch had been right. There was no bone duplication.

But a single individual of what? On preliminary examination it was very hard to say, for nothing quite like it had ever been discovered. The camp was rocking with excitement. That first night we never went to bed at all. We talked and talked. We drank beer after beer. There was a tape recorder in the camp, and a tape of the Beatles song "Lucy in the Sky with Diamonds" went belting out into the night sky, and was played at full volume over and over again out of sheer exuberance. At some point during that unforgettable evening—I no longer remember exactly when—the new fossil picked up the name of Lucy, and has been so known ever since, although its proper name— its acquisition number in the Hadar collection—is AL 288-1.

Donald Johanson and Maitland Edey
Lucy: The Beginnings of Humankind
1981

Australopithecus Ramidus

The oldest hominid found to date was discovered by a team led by Professor Tim White of the University of California at Aramis, Ethiopia, in 1992. The seventeen fossils, which date back as far as 4.4 million years ago, are judged to belong to a new species—Australopithecus ramidus. *From the evidence currently available, this species seems to represent the missing link that brings together the family trees of apes and humans.*

The earliest known member of the human family has been discovered in Ethiopia. The hominid, who lived 4.4 million years ago, is the most ape-like human ancestor yet found, according to the scientists who found it.

Tim White of the University of California at Berkeley, Gen Suwa of the University of Tokyo and Berhane Asfaw of the Ethiopian government's Paleanthropology Laboratory, found fossils of the hominid in a barren part of northern Ethiopia called the Middle Awash in the Afar Depression…. They have named it *Australopithecus ramidus* after a word meaning "root" in the language of the Afar people. The new species narrows the gap between the last common ancestor of African apes and humans and the earliest known hominids, or australopithecines.

White and his colleagues believe that the new Ethiopian species is distinct from the famous Lucy and her kin—the early hominids called *Australopithecus afarensis*. Until the latest discoveries, these were the most ancient human ancestors known. *A. afarensis* dates from about 3.9 to 2.9 million years ago, Lucy herself having lived around 3.1 million years ago. In the winters of 1992 and 1993, White and his colleagues, searching near the village of Aramis a few kilometers west of the Awash River, found fragmentary fossil specimens of the new hominid representing about 17 individuals—pieces of skull, a child's lower jawbone, many teeth and broken arm bones, three of which were from the same limb. The site is about 75 kilometers [47 miles] to the southwest of Hadar, where Lucy and several other *A. afarensis* fossils were discovered.

No one is surprised at finding a hominid older than 4 million years, and fossil hunters have long searched for specimens in the Rift Valley of East Africa. The human and African ape lineages diverged between 8 million and 6 million years ago, but no one has ever found fossils of creatures resembling apes or humans from that period in Africa. The new fossils, together with molecular evidence, place the split near to 6 million years ago.

A few nondescript fossils from four sites in Kenya are between 5.5 and 4 million years old. They are probably from australopithecines but have not proved very helpful in clarifying when certain human features first appeared. In particular, it is not known whether these creatures walked on two legs. No hip or leg bones were found among the *A. ramidus* fossils, so it is not clear whether they were bipedal either. However, there is one clue. In *A. ramidus*, the foramen magnum, the opening at the base of the skull through which the brain connects to the spinal cord, is further forward than it is in apes. This suggests that the head was balanced on the backbone. White and his colleagues claim that this feature, together with several other characteristics, including the size and shape of the teeth, distinguish the Aramis hominid from extinct and living African apes.

Not much can be said about the general appearance or lifestyle of *A. ramidus*. Since no facial bones were found at the site, a skull cannot be reconstructed. The arm bones suggest that the hominids were taller than the diminutive Lucy, who was only 105 centimeters [41 inches] in height, but shorter than some other *A. afarensis* individuals, who stood around 150 centimeters [59 inches] tall.

More is known about the environment in which the Aramis hominids lived. It was a flat plain covered with woods and forest, and according to [scientists at] Los Alamos National Laboratory in New Mexico, *A. ramidus* lived among colobus monkeys, kudus and other treeloving animals.… The forested environment may have meant that the hominids spent a good deal of time climbing about in trees, just as *A. afarensis* may have done.

The suggestion that the Aramis hominids walked on two legs reinforces the idea that open grasslands were not necessary for the evolution of bipedalism, and lends support to a theory…that an upright posture evolved for the purpose of gathering fruit from trees in open forest and woodland.…

Until more fossils of *A. ramidus* are found, and in particular more skull and lower limb bones, it is difficult to assess the relationship of this species to *A. afarensis* and to later humans. However, it is possible to speculate about two main hypotheses. One is that *A. ramidus* was the ancestor of *A. afarensis*, and that *A. afarensis* was the last common ancestor of the heavy-jawed "robust" australopithecines, such as *Paranthropus* (or *Australopithecus*) *boisei*, and of a line that eventually led to our own genus, *Homo*.

Another possibility is that *A. ramidus* was the last common ancestor of the *Homo* lineage and of the robust australopithecines. In that scheme, *A. afarensis* would not be a direct human ancestor. The robust australopithecines became extinct a little before a million years ago. A third possibility, tentatively suggested by Bernard Wood of the University of Liverpool, is that *A. ramidus* was the ancestor only of the robust australopithecines.… This would make *A. afarensis* a direct human ancestor, but not *A. ramidus*.

Sarah Bunney
New Scientist, 1 October 1994

Where Did Humans Come From?

Two rival camps hold diametrically opposed views on the origins of modern humans. The multiregional model, or the theory of regional continuity, and the population-replacement model put two different interpretations on the same fossil evidence.

Today there are two main competing scientific camps, each believing it holds the solution. Both accept that there was a migration out of Africa by *Homo erectus* populations beginning around 1 million years ago ("Out of Africa 1" as we shall call it). One camp, however, argues that there was at least one other major wave of migration ("Out of Africa 2") around 100,000 years ago, this time of anatomically modern humans—*Homo sapiens*—people who had evolved in Africa from *Homo erectus* stock and subsequently replaced all other populations in the world including the Neanderthals. Against this model of *population replacement*, the rival camp sets its model of *regional continuity*. For the followers of this latter school, there was no pronounced Out of Africa 2 migration. Instead, modern humans evolved semi-independently in different regions of the world from independent populations of Ancients (Neanderthals in Eurasia, *Homo erectus* in China and Java), with continual gene flow or interbreeding between geographically contiguous groups so that a single but racially diverse modern human species was the result.

It becomes clear that the Neanderthals —for whom we have a wealth of evidence greater than for any of our other fossil relatives—are central to this argument. Did they evolve into people like us, as the multiregionalists would have us believe, or were they an evolutionary dead end, as the proponents of population replacement would argue?…

According to current scientific thinking, speciation—the process by which new species are formed—is most commonly a product of the geographic isolation of an interbreeding group or population. Set out by Ernest Mayr, this geographical model of speciation is known as allopatry, and in the case of human evolution may draw on genetic, anatomical and archaeological evidence. Isolation can be produced either by geographical barriers, such as mountain formation or a rise in sea levels, or by new behavioural or morphological obstacles to interbreeding within a previously continuous population. The multiregional and replacement models for speciation disagree over the extent of isolation present in widely dispersed early human populations.

Multiregional evolution emphasizes continuity in both time and space. According to this model, isolation was never sufficient to allow allopatric speciation, since genes (the basic units of heredity) were circulated and exchanged between all the human populations of the Pleistocene. There could be no speciation because throughout the last 1 million years there was really only one species: *Homo sapiens*. This judgment implies that since the first dispersal of hominids out of Africa a million years or more ago, all the observable variation is within this one species. Multiregionalists argue that the mechanism of change was predominantly behavioural, with anatomy eventually evolving to accommodate progressive changes in behaviour that usually involved improvements in technology. These changes, like the genes, circulated around the inhabited world. The different regional lineages responded in similar ways to these universal forces, directing change globally towards modern-looking humans. Nevertheless, certain local differences were, at the same time, being maintained. Selection for specific features in particular environments kept them in local populations as they gradually became more modern, e.g. the large noses of Neanderthals were maintained throughout the transition to modern Europeans, probably in response to the European climate, and the strong cheek bones of Javanese *Homo erectus* were maintained in the transition to modern native Australians, perhaps due to behavioural or dietary factors. The mechanism of interregional gene flow is all-important in multiregional evolution, to continually introduce new characteristics which can be worked on by local selection, and to counterbalance the tendency to local specializations which would increase divergence between geographically remote populations.

The population replacement camp has not so far produced a comparable theoretical dogma to account for evolutionary change.... The differences between the Neanderthals and modern humans...lay in their society and culture as well as in their anatomy.... The two communities were supported by different capacities for communication—verbal, visual and symbolic—and...this in turn affected their organization of camp-sites, their exploitation of the landscape, and their colonization of new habitats. But to conclude that the Neanderthals were different from us is not to condemn them in the same way that earlier popularizers and scientists did.... The Neanderthals were not ape-men, nor missing links—they were as human as us, but they represented a different brand of humanity, one with a distinctive blend of primitive and advanced characteristics. There was nothing inevitable about the triumph of the Moderns, and a twist of Pleistocene fate could have left the Neanderthals occupying Europe to this day. The 30,000 years by which we have missed them represent only a few ticks of the Ice Age clock.

Christopher Stringer and Clive Gamble
In Search of the Neanderthals
1993

Glossary

Acheulean Lower Paleolithic culture dating to 2 million to 100,000 years ago. It owes its name to the hamlet of St. Acheul in northern France. The characteristic stone tools include numerous pieces worked on both sides, known as bifaces.

Aurignacian Upper Paleolithic culture dating to between 33,000 and 26,000 BC.

Australopithecus A fossil hominid found in eastern and southern Africa about 5 to 1 million years ago. This small biped, which ranged in height from 3 to 5 feet, had a small brain (450 to 550 cc, or about 27 to 34 cubic inches) and a massive face with projecting jaws. There were two forms: the "gracile" (*ramidus, afarensis, africanus*) and the "robust" (*aethiopicus, robustus, boisei*).

Back This term is used in the typology of stone tools to designate an abrupt retouch that has removed the cutting edge of a blade (backed blade) or bladelet (backed bladelet).

Backed tools Tools with a lateral surface that cuts roughly perpendicularly both sides of a flake, blade, or bladelet in the direction of its greatest size.

Biface Stone tool worked on both faces.

Bifacial retouch Retouch on both sides of a flint tool.

Bipedalism The ability to walk on two legs.

Blade Elongated flake with two parallel edges, whose length is more than twice its width. Blades are particularly numerous in the Upper Paleolithic traditions. Bladelets are small blades less than half an inch wide.

Borer Tool of bone, ivory, or antler with a single beveled base.

Burin Worked stone tool very common in the Upper Paleolithic. It has a beveled edge formed by the scar made by removing a bladelet and either the scar from the removal of another bladelet (dihedral burin) or a series of continuous retouches (burin on truncation).

Carbon 14 (C14) The use of this radioactive isotope of carbon to date ancient materials was discovered in 1947 by the American chemist Willard Libby. All organic matter contains carbon, including a minuscule amount of C14. When an organism dies the C14 in it starts to decay at a very slow, but consistent, rate; its half-life is 5730 years. Measuring the radio-activity remaining in a sample enables scientists to tell how long ago the organism died.

Ceraunia or thunderbolt-stone From the Greek *keraunos,* meaning lightning. Before the 19th century, this word was used to designate prehistoric tools of stone (especially polished stone), whose origin was then unknown.

Châtelperronian Culture that marks the transition between the Middle and Upper Paleolithic, around 34,000 to 30,000 BC.

Core Block of stone worked to produce the flakes, blades, and bladelets needed to make tools and weapons.

Cro-Magnons See *Homo sapiens sapiens.*

Dating (conventions) BP means "before present," the "present" being 1950, the year considered to be zero in this dating system. BC and AD have their usual meaning. Dates BC are 1950 years less than dates BP.

End-scraper Worked stone tool, very common in the Upper Paleolithic, characterized by a series of continuous retouches, forming a more or less rounded edge at the extremity of the piece.

Eocene The second part of the Tertiary period. It lasted about 20 million years, from about 58 to 27 million years ago.

Epipaleolithic Group of cultures starting at the end of the last glaciation, around 10,000 years ago, and ending in the Neolithic, around 3000 BC. This term is especially used to designate the cultures of hunter-fisher-gatherers that emerged from the Paleolithic tradition. Mesolithic cultures were more inclined towards food production and settling, but still derived much of what they needed from hunting.

Flake Fragment of rock intentionally struck from a block of raw material or from a core, by percussion or pressure. A flake has an upper face, with the scars left by earlier removals; a lower face or struck face, with the bulb of percussion; and a heel, which is part of the striking platform carried away with the flake.

Gravette point Worked stone knife with a straight back and a sharp end. This tool characterizes the Upper Paleolithic Gravettian culture and owes its name to the site of La Gravette in the Dordogne, France.

Gravettian The ubiquitous presence of Gravette points or microgravettes in all phases of the Upper Perigordian culture has led to this term being supplanted by "Perigordian."

Hominid A term for an early form of human being that does not specify gender or time.

Homo erectus This human species, which appeared in East Africa 1.7 million years ago, eventually inhabited the Old World—North Africa, Asia, and probably Europe—for almost 1.5 million years. In terms of height—*Homo erectus* could reach five and a half feet—these individuals were close to modern humans, though their cranial capacity was much smaller than ours today, varying from about 775 to 1250 cc (about 47 to 76 cubic inches). Their teeth were enormous in comparison with those of *Homo sapiens*, but they were smaller than those of earlier hominids. These hunter-gatherers were the first to control fire and set up dwellings. They invented a new technique of working stone.

Homo habilis The name of this species means skillful person (because of the ability to work tools). This is the oldest representative—as far as we are currently aware—of the genus *Homo*, which appeared in East Africa 1.8 million years ago. These individuals differed from their predecessors the australopithecines in several ways—they had a larger cranial capacity (650 to 800 cc, or 40 to 50 cubic inches), a more compact face, and smaller teeth.

Homo sapiens With a name from the Latin meaning intelligent or wise person, this most recent evolutionary form of humans developed from *Homo erectus*. These individuals show signs of evolutionary change in their increased brain size (1400 cc, or about 85 cubic inches, on average), developed frontal lobes, reduced face projection, decrease in teeth size, and the appearance of a chin.

Homo sapiens neanderthalensis An archaic subspecies of *Homo sapiens*, found mainly in Europe and

parts of Asia, appeared about 230,000 years ago and disappeared about 35,000 years ago. The Neanderthals had to adapt to very varied climate conditions—sometimes cold and dry, sometimes mild and humid—and lived in open-air camps or rock shelters, the steppes or the tundra. Physically, they were small and muscular. They had a broad, receding head with an attenuated rear (bun) and a face distinguished by a prominent nose, receding cheekbones, a low brow, and no chin. Their cranial capacity often exceeded the average of modern humans. The Neanderthals acquired a degree of sophistication, which is reflected in their tools, their spiritual life, and their hunting methods.

Homo sapiens sapiens This more evolved lineage goes back more than 100,000 years. These anatomically modern humans originated in Africa. They appear to have invaded western Europe from the East and replaced the Neanderthals. They differ from *Homo erectus* in having a larger skull with, on average, a cranial capacity of 1350 cc (about 82 cubic inches), a high brow, prominent cheekbones, and a distinct chin. The best-known example of western *Homo sapiens sapiens* is the Cro-Magnon people, who appeared in Europe 35,000 years ago. One of the most striking

features of these hunter-gatherers is the development of aesthetic feelings, closely linked to religion or magic, found in their cave art—paintings, engravings, sculptures, and ornaments.

Levallois technique Method of knapping stone which enables one to obtain large flakes of predetermined shape. The core is first shaped by a series of removals made by blows almost perpendicular to its principal plane; a second series of removals, made by tangential blows on the ridges separating the scars from the first series, gives the core a turtle-shell shape. A striking platform created at one of the core's narrow ends makes it possible to extract a single large flake. This thin, oval flake bears on one of its faces the convergent flake-scars of the second series of blows.

Magdalenian Culture of the end of the Upper Paleolithic (about 17,000 to 10,000 BC) which owes its name to the site of La Madeleine. The stone industry, based on blade production, became miniaturized towards the end of this period. The tradition of working on bone was developed, and bone pieces were often decorated and became portable art objects. Also, the zenith of cave art was in this period.

Microlith Very small piece of worked stone, usually of flint. Average length: about 1 inch.

Mousterian Prehistoric

culture that developed in the Middle Paleolithic, between about 100,000 and 35,000 years ago. It owes its name to the rock shelter of Le Moustier in the Dordogne, France. Neanderthals were associated with the Mousterian tradition. Tools were shaped from flakes that had been retouched into different forms, including side-scrapers.

Neanderthal Named after the site of Neanderthal, near Düsseldorf, Germany, where the remains of a fossil human were discovered in 1856.

Neanderthals See *Homo sapiens neanderthalensis*.

Neolithic Period of prehistory following the Paleolithic around 8000 years ago. It is characterized by polished stone tools, the making and firing of ceramics, and an important change in the way of life: People settled down, became farmers, and grouped their dwellings into villages.

Paleolithic (Old Stone Age) Period characterized by flaked stone tools during which primitive people emerged. The Paleolithic is subdivided into three great parts: the Lower Paleolithic which started about 2 million years ago and lasted until about 100,000 years ago and corresponds to the development of the Acheulean industries; the Middle Paleolithic, from about 100,000 to 35,000, during which the Neanderthals made the Mousterian industries; and the Upper Paleolithic, from about 35,000 to 10,000, during which modern humans discovered art and made great use of blade production and bone working.

Palynology The study of pollen. The fossil pollen in the sediments filling a site provides information on the plant environment at the time when the sediments were deposited. The variations in the environment reflect the evolution of the climate.

Parietal From the Latin *paries*, wall. Designates engravings, drawings, or paintings produced on the rocky walls of prehistoric caves.

Perforated baton Object of reindeer antler pierced by one or, more rarely, several quite large holes. This instrument is often decorated and is found in all Upper Paleolithic levels from the Aurignacian period on.

Perigordian Culture of the early Upper Paleolithic subdivided into the Lower Perigordian (Châtelperronian) and Upper Perigordian (Gravettian).

Pithecanthropus From the Greek *pithekos* (ape) and *anthropos* (man). Fossil of the species *Homo erectus*, discovered in 1891 in Java.

Side-scraper Stone tool, generally made on a flake, with continuous retouch on one edge. The retouched edges of side-scrapers were most usually made on one of the long sides.

Sinanthropus Fossil of *Homo erectus* discovered in 1927 at Zhoukoudian cave near Peking (Beijing).

Solutrean Upper Paleolithic culture dating to between 20,000 and 16,000 BC, which occupied a large part of western Europe. The stone industry is characterized in particular by leaf-shaped points with parallel oblique retouches that sometimes affect both sides (laurel-leaf points).

Spearthrower Instrument sculpted from reindeer antler with a hook at one end. These pieces, very characteristic of the Magdalenian culture, are often magnificent art objects. Similar objects are still used by Eskimos and Australian aborigines.

Steatopygia Major development of fatty tissue in the buttock area. This racial feature is much more marked in women than in men. It is often mentioned in relation to prehistoric "Venuses."

Stratigraphy The study of the succession of sedimentary deposits that are generally grouped into layers, taking into account their nature and content.

Tertiary Period in the Cenozoic era—the current geological era —during which mammals became dominant. It comprises the Paleocene, Eocene, Oligocene, Miocene, and Pliocene eras.

Typology Study of the forms of the stone and bone tools found in archaeological layers. This study enables one to define types, classify the tools, and produce a quantitative analysis of prehistoric industries.

Venus Name given to prehistoric statuettes, bas-reliefs, and engravings representing women.

Würm Last ice age, which stretched from about 110,000 to 10,000 years ago. It owes its name to a tributary of the Danube. It includes a period of maximum cold 18,000 years ago, during which the ice was almost 2 miles thick in places and the sea level dropped by almost 400 feet.

Glossary based on
Art et Civilisation de la Préhistoire, 1984, and
Dossier d'Archéologie,
January 1991

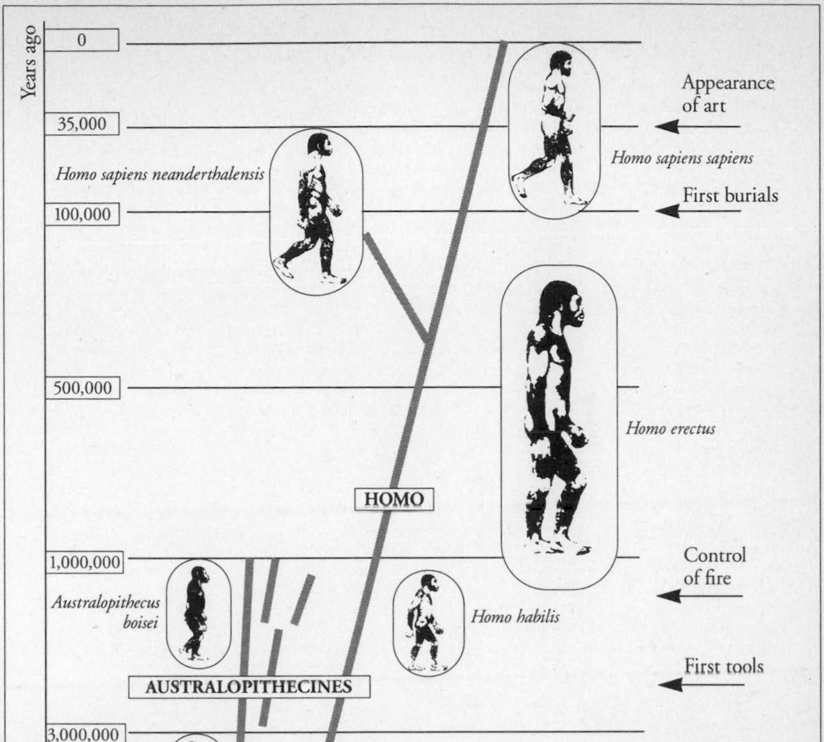

Years ago

0

35,000 — Appearance of art

Homo sapiens neanderthalensis

Homo sapiens sapiens

100,000 — First burials

500,000

Homo erectus

HOMO

1,000,000 — Control of fire

Australopithecus boisei

Homo habilis

AUSTRALOPITHECINES

First tools

3,000,000

Australopithecus afarensis

Bipedal hominids

Australopithecus ramidus

PRE-AUSTRALOPITHECINES

6,000,000

KENYAPITHECINES

Kenyapithecus

Simplified chart showing the different stages in human evolution. Outside the human lineage (Homo habilis, Homo erectus, Homo sapiens) the exact relationships between ancestors and their descendants remain problematic.

15,000,000

Further Reading

Aiello, Leslie, and Christopher Dean, eds., *An Introduction to Human Evolutionary Anatomy*, Academic Press, San Diego, 1990

Bahn, Paul G., and Jean Vertut, *Images of the Ice Age*, Facts on File, New York, 1989

Burenhult, Göran, and David H. Thomas, eds., *The First Humans: Human Origins and History to 10,000 BC*, Harper San Francisco, 1993

Cole, Sonia, *Leakey's Luck: The Life of Louis Leakey, 1903–72*, Harcourt Brace Jovanovich, New York, 1975

Eldredge, Niles, and Ian Tattersall, *The Myths of Human Evolution*,

Columbia University Press, New York, 1984

Fagan, Brian M., *The Journey from Eden: The Peopling of Our World*, Thames and Hudson, New York, 1990

Falk, Dean, *Braindance: New Discoveries about Human Brain Evolution*, Henry Holt, New York, 1994

Fleagle, John G., *Primate Adaptation and Evolution*, Academic Press, San Diego, 1988

Grine, Frederick E., ed., *Evolutionary History of the "Robust" Australopithecines*, Aldine de Gruyter, 1988

Johanson, Donald C., and Maitland A. Edey, *Lucy: The Beginnings of Humankind*, Simon

and Schuster, New York, 1981

Leakey, Mary D., *Disclosing the Past*, Doubleday, Garden City, N.Y., 1984

Leakey, Richard E., *The Making of Mankind*, Dutton, New York, 1981

Lewin, Roger, *Human Evolution*, Blackwell Scientific Publications, Cambridge, Mass., 1993

Reader, John, *Missing Links: The Hunt for Earliest Man*, Viking-Penguin, New York, 1994

Spencer, Frank, *Piltdown: A Scientific Forgery*, Oxford University Press, New York, 1990

Stringer, Christopher, and Clive Gamble, *In Search

of the Neanderthals*, Thames and Hudson, New York, 1993

Sutcliffe, Antony J., *On the Track of Ice Age Mammals*, Harvard University Press, Cambridge, Mass., 1985

Tattersall, Ian, et al., eds., *Encyclopedia of Human Evolution and Prehistory*, Garland Publishing, New York, 1988

Tobias, Philip V., ed., *Hominid Evolution: Past, Present and Future*, John Wiley and Sons, New York, 1985

Turnbaugh, William A., et al., eds., *Understanding Physical Anthropology and Archeology*, West Publishing Co., St. Paul, Minn., 1993

List of Illustrations

Index

Photograph Credits

A.F.P., Paris 85. All rights reserved 44, 129, 130, 137. Anthropological Museum, Institute of Human Paleontology, Beijing 87. Archiv für Kunst und Geschichte, Berlin 32. Archives Gallimard 14, 18, 32–3, 33, 41, 46l, 60l, 78 (4), 117b. Bibliothèque, Muséum National d'Histoire Naturelle, Paris 26a, 42a. Bibliothèque Nationale, Paris back cover, 17, 20–1, 23l, 24, 27a, 27b, 28a, 29a, 31b, 40–1, 47r, 48 inset, 48, 49a, 50–1, 51c. Bridgeman Art Library, London 34–5, 37. Bulloz, Paris/Musée Carnavalet, Paris 31a. Mario Caselli, Florence spine, 1, 13, 75, 90–1, 112–3. Charmet, Paris 19, 20, 21, 30a, 36, 43. City Art Galleries, Manchester 16a. C.N.R.S. 83a, 83b. C.N.R.S.-L.M.C. 59a. C.N.R.S.-S.M.C. 58r. Cosmos, Paris 110. Cosmos, Paris/Earth Satellite Corporation/S.P.L. 127. Cosmos, Paris/E. Ferobelli 58l. Cosmos, Paris/J. Reader/S.P.L. 35, 46r, 52, 100–1. Cosmos, Paris/L. Pesek/S.P.L. 62–3. Cosmos, Paris/S.P.L. 56, 59b, 65, 66–7, 72l, 84a, 84l, 92b. Dagli Orti, Paris 15, 23r, 38, 38–9, 44–5. DITE, Paris 82. Edimédia, Paris 104. Explorer, Paris/S. Cordier 64–5; Fiore 71r; A. Bertrand 72a; Lorne 112. Frisano, Paris 72–3. Claus Hausman, Munich 109a. Institut de Paléontologie, Paris/Serette 61b. Jacana, Vanves/J.M. Labat 57, 60r. David Keith Jones, Lichfield, England 71l, 88. Courtesy Michael Joseph Ltd., London 140. Hubert Josse, Paris 22a, 106–7. Keystone, Paris 126. Kharbine Tapabor, Paris 42b. Lauros-Giraudon, Vanves 18–9. Collection Prof. Henry de Lumley, Laboratoire de Préhistoire, Institut de Paléontologie Humaine, Paris 79 (6, 7), 88–9, 90, 94–5. Magnum, Paris/E. Lessing 93a, 115. Daniel Moignot, Paris 74–5, 92a. Musée Boucher de Perthes, Abbeville 29b. Musée de l'Homme, Paris 28b, 52–3, 70r, 80 (9), 81 (17), 94a, 94b, 117a, 117r, 121, 122–3, 151; D. Destable 40, 108br; M. Delaplanche 45, 49bl; J. Oster 76a, 80 (10, 11), 89, 105, 108l, 109bl, 109br, 111r, 124–5; D. Ponsard 80 (12). Musée du Périgord, Périgueux 49br. Muséum National d'Histoire Naturelle, Paris/D. Serette 22b; B. Fayes 25. NASA 39. Nationaal Natuurhistorisch Museum, Leiden 55a. National Geographic Society, Washington, D.C. 78 (1, 2); David Brill 64b, 70l; Joseph J. Scherschel 69; Melville Bell Grosvenor 77; Robert F. Sisson 96–7, 97; David S. Boyer 98; Des Barlett 99a, 99b. Natural History Museum, London 55b. Novosti, Paris 118b. Oronoz, Madrid 16b. Pavillion/Institut Anthropos, Moravské Muzeum, Brno, Czech Republic 86, 106, 114, 128. Réunion des Musées Nationaux, Paris 30c, 116, 119a, 120a. Rijksmuseum voor Volkenkunde, Leiden 53. H. Roche, Laboratoire de Paléontologie 76b. Roger Viollet, Paris 54r. Société Historique et Archéologique du Périgord 47l. Fondation Teilhard de Chardin, Paris 93b. Herbert Thomas, Paris 78 (3), 79 (5), 81 (15), 84br, 102a, 102b, 103a, 103c, 103b, 142. Gilles Tosello, Paris 2–3, 4–5, 6–7, 8–9, 10–1, 68. Ullstein, Berlin 51a. B. Vandermeersch, Bordeaux 111l. N. K.Vereshchagin 118a. J. Vertut, Issy-les-Moulineaux 79 (8), 81 (13, 14, 16, 18), 108ar, 109bc, 119b, 120b. Jean Vertut Archives front cover. J. Vigne, Gennevilliers 54a, 61a. The Dean and Chapter of Westminster 26b.

Text Credits

Grateful acknowledgment is made for use of material from the following works: Sarah Bunney, *New Scientist*, 1 October 1994. Reprinted by permission of *New Scientist*, London (pp. 144–5); Professor Raymond A. Dart, *Nature*, 7 February 1925. Reprinted by permission from *Nature*, Vol. 115, pp. 195–9, copyright © 1925 Macmillan Magazines Ltd (pp. 136–8); Ray Ginger, *Six Days or Forever?*, Oxford University Press, 1974. Reprinted by permission of Victoria Brandon (pp. 134–5); Donald C. Johanson and Maitland A. Edey, *Lucy: The Beginnings of Humankind*, Granada Publishing, 1981. Reprinted by permission of the Peters Fraser & Dunlop Group Ltd, London (pp. 142–3); Sir Arthur Keith, *The British Medical Journal*, 14 February 1925. Reprinted by permission of BMJ Publishing Group, London (pp. 138–9); Richard E. Leakey, *One Life: Richard E. Leakey: An Autobiography*, Michael Joseph Ltd, London, 1983, pp. 147, 148–9, 153–4, copyright © Sherma BV 1983. Reproduced by permission of Richard E. Leakey (pp. 140–1); Christopher Stringer and Clive Gamble, *In Search of the Neanderthals*, Thames and Hudson Ltd, 1993. Reprinted by permission of the authors (pp. 146–7)

Herbert Thomas, deputy director of the
paleoanthropology and prehistory laboratory
at the Collège de France, has directed numerous
excavations in Africa and Asia, for which he has been
given an award by the Fondation de la Vocation and the
Tchihatchef prize by the Académie des Sciences. A member
of the Explorers Club, he is the author of more than
a hundred scientific and popular works. In 1974 he took part
in the Franco-American expedition to Afar, Ethiopia,
which led to the discovery of Lucy. Since 1987 he has been
leader of the Franco-Omani paleontological expedition,
on which the oldest anthropoid primates have been
discovered. Among other major projects, he is leading an
expedition to Vietnam under the aegis of the Ushuaïa
Foundation, in search of further evidence of the sao la,
a large mammal that was discovered in 1992.

For Alexandre

Translated from the French by Paul G. Bahn

For Harry N. Abrams, Inc.
Editor: Sharon AvRutick
Typographic Designer: Elissa Ichiyasu
Design Supervisor: Miko McGinty
Assistant Designer: Tina Thompson
Text Permissions: Neil Ryder Hoos

Library of Congress Catalog Card Number: 95–75661

ISBN 0–8109–2866–3

© Gallimard 1994

English translation copyright © 1995 Harry N. Abrams, Inc., New York,
and Thames and Hudson Ltd., London

Published in 1995 by Harry N. Abrams, Inc., New York
A Times Mirror Company

Printed and bound in Italy by Editoriale Libraria, Trieste